# Praise for *Hoofprints on the Land*

'Ilse's deep understanding of herding cultures, and their relationship with the land and life itself, is both moving and revelatory. Pastoralism, she shows us brilliantly, is not a marginal issue but a symbiotic partnership between animals, humans and ecosystems that should be at the heart of our efforts to heal the planet. I loved this book.'

Isabella Tree, author of Wilding

'Ilse Köhler-Rollefson's *Hoofprints on the Land* reminds us that animals are not objects to be manipulated in factory farms. They are not a "technology" to be pushed to obsolescence and extinction in the new rush for making fake milk, fake cheese and fake meat. Ilse shows how animals are sentient beings, subjects not objects, members of our families. Animals should never have been put in factory farms. Factory farms violate the rights of animals and contribute to pollution, including climate change. Ilse shows that free-range animals and animals in pastoral cultures are a solution to climate change that factory farming has contributed to.'

Vandana Shiva, author of Terra Viva

'Grazing done right can improve biodiversity and regenerate pastureland. You will gain many insights into how to improve land from *Hoofprints on the Land*.'

Temple Grandin,<br>author of Animals in Translation

'I never knew how fascinating a book about herding and grazing could be, never understood how vital is the part that pastoralists play concerning the health of the planet and its grazing animals. But I have drunk the delicious camel milk in Ilse Köhler-Rollefson's dairy, and am a convert to everything she espouses. This book is remarkable: scholarly, accessible and hugely important.'

Joanna Lumley

'A beautiful, deeply thoughtful and intelligent book that completely reframes the fraught discussion around the role of animals in our food system. Every reader will not only learn a great deal but will also see the world in a new and better light.'

Nicolette Hahn Niman, author of *Defending Beef*

'Entirely timely, unique and massively thought provoking. It raises a whole host of intriguing issues which often, in my view, although identified as pertinent to the Southern Hemisphere, have clear and painful parallels in the north. I am not sure that many involved with limited appreciations of how livestock farming works will realise these synergies. But they should be illuminated and understood. Ilse's depth of knowledge of subject is splendid.'

Derek Gow, author of *Bringing Back the Beaver*

'Inspiration for western agriculture as an extension of the ever-growing interest in regenerative agriculture, *Hoofprints on the Land* opens our minds to the important role nomadic herding could play in securing the future of people in dry lands, while also playing a vital role in environmental management. For most of us farming in temperate climes, nomadism may seem an irrelevance, a nostalgia from bygone ages; *Hoofprints on the Land* helps us to understand how misguided these impressions are.'

Helen Browning,
chief executive, Soil Association

'A must-read for anyone who cares about the Earth. *Hoofprints on the Land* is a powerful story of hope, sharing a way of producing food that gives back more than it takes away from nature and humanity combined. A genuinely inspirational book – I absolutely devoured it.'

Lynn Cassells,
coauthor of *Our Wild Farming Life*

'In *Hoofprints on the Land*, Ilse Köhler-Rollefson shows how traditional herding cultures today, often impoverished and overlooked, might still save the planet. This is a passionate, important book, a must-read for anyone interested in ecology or food or our future coexistence with wild and domestic animals.'

Brad Kessler, author of *Goat Song*

'A provocative and thoughtful meditation on the necessity of distinguishing between industrialised farming and traditional methods of pastoralism when discussing food security and the future of agriculture. Transhumance has been around since the beginning of animal domestication and works within established ecosystems, putting in more than it takes out. There is wisdom in age-old practices of animal herding that deserve to be preserved and protected.'

DR ROSS BARNETT, author of *The Missing Lynx*

'Pastoralists care for the Earth, provide a flood of protein resources, and maintain cultures of enormous depth. All of this is stunningly clear from the story of the Raika camel herders of Rajasthan, told by one of their closest allies and most thoughtful observers. This wonderfully documented book shows that herding is a twenty-first-century technology for sustainability.'

PAUL ROBBINS, dean, Nelson Institute for Environmental Studies, University of Wisconsin–Madison

'All of us who are concerned and worried about pastoralists and traditional livestock herders and their role in our world simply have to read this book by Ilse Köhler-Rollefson – and soon. Those of us who are unconcerned or unaware of how intertwined our world is with theirs, simply have to read this book – even sooner.'

P. SAINATH, author of *Everybody Loves a Good Drought*

'For centuries, the Raika communities have lived in harmony with nature, in the course of which they have developed some of India's most vibrant oral and folk traditions. Having worked with the Raika community for many decades, I believe their worldview, traditions and way of regenerative and sustainable livestock rearing show the world an important way forward in dealing with many challenges that we face today, especially in the area of climate change.'

WILLIAM NANDA BISSELL, executive vice chairman, Fabindia Limited

## Also by Ilse Köhler-Rollefson

*Camel Karma:*
*Twenty Years Among India's Camel Nomads*

*Invisible Guardians:*
*Women Manage Livestock Diversity*

*A Field Manual of Camel Diseases:*
*Traditional and Modern Healthcare for the Dromedary*

# Hoofprints on the Land

*How Traditional Herding and Grazing Can Restore the Soil and Bring Animal Agriculture Back in Balance with the Earth*

ILSE KÖHLER-ROLLEFSON

*Foreword by* FRED PROVENZA

Chelsea Green Publishing
White River Junction, Vermont
London, UK

Commissioning Editor: Muna Reyal
Project Manager: Angela Boyle
Copy Editor: Caroline West
Proofreader: Anne Sheasby
Indexer: Nancy A. Crompton
Designer: Melissa Jacobson
Page Layout: Abrah Griggs

Front cover photograph shows a Raika shepherd returning from his daily
grazing round in the forest below Kumbhalgarh Fort in Rajasthan, India.

Printed in the United Kingdom.
First printing December 2022.
10 9 8 7 6 5 4 3 2 1    22 23 24 25 26

ISBN 978-1-64502-152-0 (paperback) | ISBN 978-1-64502-153-7 (ebook) |
ISBN 978-1-64502-154-4 (audio book)

Library of Congress Cataloging-in-Publication Data is available.

Chelsea Green Publishing
85 North Main Street, Suite 120
White River Junction, Vermont USA

Somerset House
London, UK

www.chelseagreen.com

*For Justus, Lotta, Jette, and Felia,*

*Shauriya, Siya, Lakshyaraj and Kanika,*

*Kushi, Kaviya and all the children*
*fortunate to grow up with livestock*
*or have a herding heritage.*

# Contents

PART THREE

## Why Herding Is the Future

## Herd

From Middle English *herde, heerde, heorde*, from Old English *hierd, heord* (herd, flock; keeping, care, custody), from Proto-Germanic *erdō* (herd), from Proto-Indo-European *kerd* (file, row, herd).[1]

## Herding

A set of techniques, interactions, cognitive expectations that build a specific relational configuration between herds, herders and the environment.[2]

# Foreword

H*omo sapiens* have been on Earth for roughly 300,000 years. During most of that time, our ancestors hunted and gathered for nourishment. Only in the past 10,000 years did we transition from hunters and gatherers to pastoralists and small-scale farmers and ranchers. And only during the past century did we create civilisations dependent upon industrial agriculture, a move some anthropologists claim is the worst mistake our species ever made. That's in part because we transformed from sunlight-driven ecological economies linked with the landscapes that nourish and sustain us, to economies disconnected from nature and utterly dependent upon fossil fuels.

It's easy to understand why we embraced fossil fuels when you understand that the energy in a single barrel of oil is equivalent to 10 to 12 years of hard work by a fit human. Fossil fuels enabled agriculture to evolve from long days of back-breaking work into an industry where machines do most of the work. Globally, we now consume 100 million barrels of oil each day across all human activities.

Fossil fuels enabled human populations to rise precipitously from two million people in 10,000 BCE to a little less than a billion in 1800 to nearly eight billion people today. Our populations expanded exponentially during the twentieth century due to fossil fuels and industrial agriculture.

For better or worse, oil and natural gas are projected to run out by the middle of this century, coal by the end. Prices will soar as fossil fuel abundance decreases. What will become of human populations and economies as the availability of fossil fuels declines? Which peoples will suffer most due to lack of fossil fuels: Pastoralists sustained mainly by solar economies or people in nations like the United States and the United Kingdom who live almost exclusively on fossil-fuel economies? The United States population of almost 330 million used almost 20 million barrels of oil each day in 2019, and the United Kingdom's population of over 67 million used over 1 million barrels a day. And 2019 was an unrepeatable low-use year. The implications are dire, especially for the countries most dependent upon fossil fuels to support their economies and ways of life.

However, this seeming catastrophe is highlighting opportunities to produce foods in ways that nurture soil, water, plants, herbivores and people. Farming, ranching and pastoral ways of life could once again be at the heart of communities, but from soils and plants to livestock and humans, we will need to relearn what it means to be locally co-evolving with nature's communities. In the process, we will need to transition our relationships with landscapes from *ego*-logical to *eco*-logical.

That's the tale at the heart of Ilse's book as she beckons us to imagine a different model of food production. Through this wonderful account, she takes us on her personal journey. Trained as a veterinarian in Germany, she was disillusioned with her work, so she became an anthropologist and travelled to Rajasthan to study the Raika camel nomads. Her work with pastoralists, as she created a new life, took her to the international stage, highlighting the beauties and values of pastoral ways of life and defending their rights and those of shepherding cultures.

Today, livestock are under attack globally, ostensibly because they adversely affect human and environmental health, with an emphasis on the harmful effects of their emissions of the greenhouse gas methane. This perspective is reflected in the many papers by scientists from respected universities who paint rosy pictures of a future without livestock. In a vivid illustration of this mass delusion, some societies are now convincing themselves that ultra-processed plant-based faux meat and dairy is better than the real thing and that nature is a feeble-minded nitwit compared to the 'time-tested wisdom' of Silicon Valley technologies.

While emphasising issues with intensive livestock farming, Ilse's book is about those largely invisible herding cultures that regard farm animals as family rather than objects, and whose relationships with them are based not on exploitation but on reciprocity. Pastoralists engage with livestock in ways that are simultaneously good for animals, people and the planet. They are essential to upholding the web of life on Earth and ensuring its future functioning.

Ilse describes how herding cultures around the world have created an extensive body of knowledge and expertise about managing in partnership with livestock in an ethical way that perceives humans as a part of nature, rather than apart from and in antagonism with the natural communities that nourish and sustain all life. The ethics and practices of working with livestock in nature are antithetical to the view of animals as 'machines' from whom profits are to be maximised, and land as a 'commodity' from which resources are to be extracted.

The animal science profession prides itself on producing more outputs – eggs, meat, milk – using fewer animals and inputs but with little concern about the downstream costs: diminished plant and animal diversity below and above ground as monoculture crops are grown to feed livestock; appalling animal welfare; and loss of rural livelihoods. The machine model – animals as machines and genes as destiny – championed by many animal scientists has a penchant not to engage any of these 'externalities' that are caused by the 'efficient system' they promote.

This machine model is in stark contrast with the story Ilse tells of the ongoing co-evolution of pastoralists, livestock and environments. Pastoralists appreciate that genes are expressed in ever-changing landscapes and that culture is part of the process. These relationships involve extended families, and pastoralists strive to own portfolios of maternal lines that cover all eventualities.

A 'breed' for them is thus never static, ever a work-in-progress, constantly evolving as environments change. Merely seeking to conserve breeds and genes is yet another example of the reductionist approach that overlooks the importance of biological diversity as a complex phenomenon with genetic, epigenetic and cultural dimensions that link livestock with ever-changing environments.

Moreover, as Ilse points out, the industrial farming systems that produce the crops that support industrial livestock production systems are nearly always in conflict with nature. A farmer begins by clearing the land of native vegetation, and then prepares the soil by ploughing, harrowing and fertilising. The next step is to seed a monoculture crop that is then protected with several applications of chemical fertilisers, herbicides, pesticides and fungicides. Finally, the crop is harvested and brought back to the barn or to the nearest silo, where it is stored for transport to a processing unit that will turn it into ultra-processed foods. And in richer countries these activities are performed by machines that require fossil fuel. Consequently, agriculture is the largest contributor to biodiversity loss and the second largest industrial contributor to climate change.

To produce one calorie of food now requires two calories of fossil fuels for machinery, fertilisers, herbicides and insecticides. We use another eight to twelve calories to process, package, deliver, store and cook modern food. No species can survive long when expending ten to fourteen calories to obtain one calorie of energy.

Livestock and pastoralists are linked with landscapes through palates. An attuned palate, which enables creatures to meet their needs for nutrients and self-medicate to rectify maladies, evolves from three interrelated processes: flavour-feedback associations, availability of phytochemically rich foods, and learning in utero and early in life to eat nourishing combinations of foods. These processes occur when wild or domestic herbivores forage on phytochemically rich landscapes, are less common when domestic herbivores forage on monoculture pastures, are close to zero for herbivores in feedlots, and are increasingly rare for people who forage in modern food outlets.

Unlike our ancestors, the palates of many people are no longer linked in healthy ways with fertile landscapes that nourish the foods we eat. Industrial farming and selection for yield, appearance and transportability diminish the flavour, phytochemical richness and nutritive value of fruits and vegetables for humans. Phytochemically impoverished pastures and feedlot diets can adversely affect the health of livestock and the flavour and nutritive value of meat and dairy for humans.

In contrast, when livestock eat diverse mixtures of plant species, the thousands of phytochemicals that plants produce bolster their health and protect livestock against diseases and pathogens through anti-inflammatory, antimicrobial, antiparasitic and immunomodulatory effects. The benefits to humans from eating phytochemically and biochemically rich meat and dairy accrue as livestock assimilate some plant phytochemicals and convert others into metabolites, all of which become meat, fat and milk that promote human health.

While the flavours of produce, meat and dairy have become blander, processed foods become ever more desirable as the food industry learns to link synthetic flavours with feedback from energy-rich compounds that obscure nutritional sameness and diminish health. Thus, the roles plants and animals once played in human nutrition have been usurped by processed foods that are altered, fortified and enriched in ways that can adversely affect appetitive states and food preferences. The need to amend foods would be eliminated by naturally growing phytochemically rich fruits and vegetables, by allowing livestock to forage on phytochemically rich diets, and by creating cultures that know how to combine foods into meals that nourish and satiate.

Indeed, pastoralists are models for how to nourish people, the animals in their care and the diversity of plant and animal life on Earth. They raise and steward resilient breeds in diverse natural environments of immense value, now and in the future, even as we are all increasingly challenged to adapt to higher temperatures and less predictable weather patterns. The ability to adapt cannot be achieved without people who possess the skills and dedication to survive in challenging environments.

Pastoral knowledge will thus be essential as we transition from fossil-fuel based economies to the sun-driven economies that have sustained life for millennia. Beyond these important concerns are perhaps even greater considerations, as Ilse reminds us: 'When you are in trouble, you do not abandon long-standing relationships, you nurture them back to health.' And as the Global Gathering of Women Pastoralists concludes in the Mera Declaration, '...it is by remaining pastoralists that we can be of greatest service to the entire human community.'

Dr Fred Provenza, professor emeritus,
behavioral ecology, Department of Wildland Resources,
Utah State University; author of *Nourishment*

## INTRODUCTION

# 'Our animals are like our children'

*The sheep are like our parents.*

*Diné* Native Americans, USA[1]

### FEBRUARY 1991

It was the time in the morning when the cattle were assembling in the centre of the village waiting for the *gual*, their communal herdsman, to take them out to pasture. We parked our car next to a tea stall, then Dr Dewaram led us into a maze of lanes that were bounded by man-high thorn fences on both sides. Peacocks called mournfully in the background as we dodged placid grey cows with intimidating horns who strode purposefully towards the village square to join their herd.

'This is the Raika *dhani*,' Dr Dewaram explained, looking at me over his shoulder. 'The part of the village where only the Raika live and where there is space for them to keep their animals when they come back from grazing at night.'

Before I could ask any more questions, Dr Dewaram, Vinay and I veered into an opening in the spiky palisades of scrub and faced a hut covered with crooked roof tiles and shaded by a monumental neem tree whose arrowhead-shaped leaves rustled gently and shimmered in the morning light. The dwelling would not have looked out of place in a fairy tale and was surrounded by a jumble of camels. Their massive, zeppelin-shaped bodies packed the small stockade almost entirely, and they seemed so entwined that it was difficult to see where one camel ended and the next one began. A three-dimensional tessellation. They were just awakening – maybe 25 of them. Some were lying on their side, dreamily and drowsily stretching out their legs; others were kneeling, holding up their heads alertly while ruminating rhythmically and systematically. A couple had arisen and turned their heads towards us with a mildly curious expression on their long faces.

I noticed a boy, probably in his early teens, holding a metal bowl in his hand, slithering around between the cluttered creatures. And then, incongruously, a toddler, wearing only a pair of dusty shorts. Not more than two years old, with a totally absorbed expression on his earnest face, he tottered about in the spaces within this conglomerate of camels, supporting himself on the protruding hip bones of the resting animals as if they were soft playground equipment. I held my breath as one of the camels lazily stretched out her knobbly kneed legs, expecting her to knock over the toddler with her massive saucer-shaped feet.

But nothing happened. I was the only one who worried or even took note. The owner, Savantiba – gaunt, glistening face, white turban, white shirt, white *dhoti* – focused all his attention on us, the guests. We were seated on a charpoy quickly retrieved from the inside of the hut. Momentarily, a few older men clad in lavish red turbans and frilly white tops, and with golden pendants around their necks and dangling from their ears, drifted in. They stepped out of their beaky, metal-studded shoes to sit in a neat row on a coarse, camel-coloured rug with black stripes that had been spread out; a clay pipe was fired up and passed around between them. Smoke spiralled up, giving them even more the flair of characters out of *One Thousand and One Nights*. I presumed they were Raika elders.

It had only been a few weeks ago that I first heard about the Raika at India's National Research Centre on Camels in Bikaner,

Rajasthan. I was there on a fellowship to study 'camel husbandry and socioeconomics' in India. The subject had sounded great when I wrote the proposal, but researching it had proven utterly frustrating. I had spent weeks trying to track down camel nomads, but they had been as elusive as a mirage. I had never even got close to them and my project felt like an utter failure, bound to sink me academically. But one day, a shy and slight man in his early thirties had walked into the research centre and was introduced to me as Dr Dewaram Dewasi, the first veterinarian from the Raika camel-breeding caste. With a very quiet and hushed voice, as if he was afraid of being overheard by the scientists, Dr Dewaram explained to me that the primal ancestor of the Raika caste had been created by Lord Shiva for the specific purpose of herding camels. As he plied me with further intriguing titbits about his people, I hung on his words and, when he offered to introduce me to some herders he knew near his posting at the Government Veterinary Hospital in Sadri, I did not need any prodding. I had taken the first opportunity to hire a stately, ivory-coloured Ambassador car in Jodhpur to motor down to Sadri which was located in Godwar, an area at the edge of the Great Indian Desert about halfway between Jodhpur and Udaipur in central Rajasthan. On arrival at the hospital the evening before, I had not only been greeted by Dr Dewaram but also by his friend, Vinay Srivastava, an anthropologist who had already been studying the community for a few months. Vinay, who would go on to a stellar career and eventually become Director of the Anthropological Survey of India, was erudite with an upper-class accent and sprinkled his talk with references to famous anthropologists and their theories. We spent a congenial evening talking about the Raika and then, very early that morning, the three of us had bumped across potholed tracks to the village of Latada.

Although I could not understand one speck of their conversation outside the Raika hut, I surmised that Vinay and Dr Dewaram were explaining why we had come, and about my interest in camels. I absorbed myself in the scene, feeling grateful that I was finally,

finally among camel herders. Although I was vaguely aware of the engaged discussion that was going on between Savantiba and my two companions, it seemed as if it was behind a curtain. There was a stillness in the air, and I felt relaxed and at home. I was spellbound by the older boy who had snuggled up to the hindquarters of one of the standing camels. Balancing on one leg and cradling the bowl on the knee of the other, he quickly and rhythmically filled it with milk, while the camel went into a form of trance, judging from its beatific expression and half-closed eyes. It seemed a meditative act, bearing little relationship to the automated extraction of milk accompanied by the clanking of heavy steel equipment that I was familiar with from my earlier work on dairy farms in Germany.

The boy disappeared with the milk into the hut. In a short while, he reappeared with a kettle filled with camel milk tea. He poured the beige brew into small metal bowls. Blowing over the hot liquid to dissipate its heat, we sipped it slowly and carefully. Between sips, Dr Dewaram started to speak:

'This family is really in a bad situation. They have lost a number of camels in the last few weeks. They have died of a disease, a disease that is new to them. They have never seen this kind of disease before in their camels, and they are very worried,' adding, after a brief and thoughtful pause, 'They hope we can help them. I have told them you are also a veterinarian.'

Vinay nodded his head in sympathy, 'They really are poor. Savantiba was saying that they have always made their living from selling male camels at the Pushkar Fair. But in the last few years it has become difficult to sell. Only a few traders turn up, so they don't get a good price for their young camels, or are not even able to sell them.'

Dewaram continued, 'Another big problem they have is with finding grazing. He said that a wildlife sanctuary has been set up in their traditional grazing places in the forest and that area is now closed to them. Therefore, the camels are often hungry and that causes them to be sick and to have miscarriages. So camel herding is a big struggle these days and it's very difficult to make a living from it.'

He looked at the boy who had served the tea, 'He is going to Bombay tonight to find a job and from there he will send money back to the family.'

'Oh, that is terrible,' I said. 'So sad.' I felt sorry for the young boy, wondering how he would fare in the big city.

'It's the normal thing to happen for Raika boys, most of them go to the cities for work.'

Vinay made it sound as if it was nothing to worry about, so I asked about the camels:

'So, what are they going to do with the herd, if the camels do not provide any income? Will they have to sell them?'

Now Vinay shook his head, sadly but firmly. 'No! No, they will definitely not sell this herd. They have a rule in their caste to never sell female camels, so they will hang on to it.'

'But why is that? How can they do that?'

Vinay looked at me as if I was a somewhat dim student. 'How can they not do that? Please remember what we discussed yesterday evening. The Raika are the guardians of the camel. It is their duty to take care of camels, given to them by Lord Shiva. They cannot just abandon their camels because they are not profitable. Theirs is a moral economy. The rule is that female camels are passed on from one generation to the next. The only exception is at the time of marriage when the bride's family gifts a female camel to the bridegroom.'

'Yes, I remember, we talked about that. But can they not make any other uses of their camels? What about meat and milk?' I prodded, though sensing that Vinay would be irritated by my naiveté.

'No, they have rules against using camels for meat and selling milk, and even their wool. We talked about that yesterday!' Vinay said sternly, so that I barely dared to raise my next question.

'But what would happen if somebody did it anyway?'

'If somebody breaks these rules, they will no longer be part of the community. They will be shunned. You have heard of outcasting? Nobody will come to their house anymore and they can no longer attend any social functions. Nobody will want to marry into such a family,' Vinay said with emphasis, while Dewaram silently nodded in agreement.

I am trained as a veterinarian and against that background it was difficult to fathom why anybody would be keeping livestock unless it was profitable, or as a rich person's hobby. From working as a large animal practitioner in Germany, I knew that decisions about the fate of a farm animal are always economic: they are given treatment only when it makes monetary sense, when the costs of the therapy are smaller than the anticipated returns. Otherwise, they go to the butcher. I wondered if Vinay and Dewaram were romanticising the Raika and was not sure what to think.

But I was hooked. After this first encounter in Latada, my research fortunes were magically reversed. Together with Dr Dewaram and Hanwant Singh, our driver, I criss-crossed the Godwar area to interview camel herders to get to the bottom of the issue and to better understand the problems outlined by Savantiba: an unknown disease, exclusion from traditional grazing areas, no market for camels. The one sentence I heard more often than any other was 'Animals are like our children'. The Raika had an intimate association not only with the camel, but also with sheep and goats and cattle. The tenor was that it would be failing their God and their *dharma*, their divine duty, if they abandoned their herds. Without the bond with camels, their identity and existence were at stake.

---

It has been more than 30 years since that first encounter with the Raika; the world has since changed and so have I. A generation of young Raika has been forced to find employment, usually of a very menial and often hazardous sort, in India's exponentially growing urban areas. The community's intimate association with the camel and other livestock is threatening to become a mere memory for its younger generation, something they hear about only from their grandparents, and even that remembrance may soon be lost. While global camel numbers have doubled in the last 50 years, Rajasthan's camel population has shrunk from over one million to less than 200,000 today. In order to save what is an iconic part of the local heritage, the government of Rajasthan declared the camel its state animal in 2014, prohibiting its export and slaughter, among other things. But what was a well-intentioned move only sped up the decline, as it rendered it virtually impossible for the Raika to earn a living from their herds. Now this exemplary culture that once cared for camels, and other livestock, is on its last legs.

I have been a close witness to this unravelling, herd by herd, family by family, village by village. Aghast, I threw myself into documenting and raising awareness about this disbanding of a caretaking culture and the abandonment of a way of life so much in tune with nature. In a rather desperate and last-ditch effort to create income for the Raika and thereby help them to save their camels, I began marketing products from living camels. It is a labour of love rather than a commercial success. After trying out camel poo paper and experimenting with camel wool, we have managed to set up India's first dedicated camel dairy. And we have accumulated a small herd of camels that the Raika begged us to buy because they could no longer afford to keep them.

Now I am standing here on the balcony of my house at the edge of the Great Indian Desert and it's not yet light. There is not even a hint of luminosity above the wavy contours of Rajasthan's Aravalli Hills, which rise to the east of me and could indicate the sun's imminent arrival. But the radiances of the half moon and the morning star suffice to illuminate the entanglement of camels that are resting/ruminating on the patch of flat land that I am looking down on. They have arranged themselves into a cluster of intertwined bodies. There is room to spread out and they could easily respect what we humans call personal space. But as social animals they usually prefer to rub shoulders during the night.

It's the end of the monsoon season. For three months I have had the privilege and joy of having our camel herd stay with us on our land. Actually, they are only here at night and in the early morning; during the day they troop into the dry and thorny scrub forest that blankets the slopes of the Aravallis to browse on a variety of vegetation, mostly trees whose various parts – flowers, pods, leaves, bark and wood – are also known for their medicinal value.

Their sojourn here will soon be over. In the next few days, they will move to the agricultural land in the west and go about their job of de-weeding and manuring farmers' fields. The herd has been impatient to leave for some time and would normally already be gone at this time of year. But one of the camels named Rathi has what seems like a neuro-muscular disorder – she has problems getting up and totters, rather than walks. So, Madhuram, our herder, has decided to wait until she is better. I am grateful for this delayed departure, as the presence of both Madhuram and the camels has a calming, relaxing effect on me. Witnessing how they put into effect the basic principles of life on Earth – caring, sharing, moving, feeding, fertilising, propagating, recycling, regenerating – distracts me from worrying about worldly issues. It grounds me.

There is Madhuram now, emerging from the shadows of a group of trees, a slight and lithe figure, with a conical milking vessel dangling from his shoulder. He circles the herd and then addresses one of them by her name – Moomal – and gives her a slap on her croup to wake her and ask her to stand up. Languidly she complies with his request, stretches to the sky and shakes herself, while her youngster, immediately alert, dives for her udder and starts suckling. Madhuram has zeroed in from the other side and, standing on one leg with the vessel perched on his raised knee, reaches out for the teats and directs the milk that spurts out into the vessel. Within less than a minute, the container is filled to the brim with a crown of foam hovering above it. While Moomal's son continues to suckle, Madhuram empties the liquid into a 10-litre (2-gallon) milk can made from steel and then repeats the procedure with Dholi, Monki and Mewari – all the ladies that have calves at foot.

In the meantime, the first rays of the sun have broken over the eastern horizon and tinge the tableau below me with a soft hue that makes nature look crisp and the world in fine fettle. One can't help but feel optimistic, I ponder, as the composition of the scene below me changes and the cluster of camels disperses into its constituents. Only Rathi and another camel, which is either lazy or just wants to keep her company, are still sitting down on the ground. The others have ventured out in search of some breakfast nibbles, clustering around the acacia trees that dot our land. Standing on tiptoes and stretching out their necks, which makes them look like giraffes, they seek previously untouched branches – they don't like to eat from used plates, so to speak – and grab twigs with their front teeth to slice off the leaves and pods. The young camels are less interested in food than their mothers, preferring to play and frolic around in a separate group.

I am absorbed in the scene when it is interrupted by the sound of a motorbike. Hariram, wearing the same outfit as Madhuram – a red turban and white shirt and *dhoti* denoting him as an active herder – has come to pick up the milk. After a brief chat with Madhuram about how to help the camel with the standing problem, he whizzes off to our dairy, from where the milk will be dispatched all over India to people with assorted health problems. Chief among them are autistic children who often respond astonishingly well to camel milk, but also cancer patients and people with diabetes.

The milk having been sent on its way and the first chore of the morning completed, it is now time for a cup of camel milk tea. Madhuram carries his milking vessel into the kitchen and starts brewing it. Shortly afterwards, three small, bright-eyed kids from the neighbourhood – two boys and a girl, all under ten years old – walk in with metal cups in their hands. They seat themselves on the floor in an orderly row, as if they were in school, and wait for Madhuram to fill their cups with camel milk. *Tatta*, good-bye, they say, waving and smiling at me. They are the children of sharecroppers who are housed in a tattered plastic tent on a harvested field close by. I regularly pass it on my way to the office and, from the plastic wrappers that litter the vicinity, my impression is that a big part of their diet are the salty and spicy snacks that dangle on the storefronts of practically every Indian village shop. During the COVID-19 pandemic, many people did not have enough to eat and so we distributed camel milk to those people who needed it and to a school, as well as to tuberculosis patients. In some cases, the effects have been almost miraculous – our TB patient, Rupa, who could no longer walk and had been given up on by doctors, is now up and about doing rather strenuous agricultural work.

After having had his tea, Madhuram goes back to the herd. With a typical trilling sound, he tells the camels to re-assemble and gradually they regroup. He checks each camel for wounds and applies some oil and ointments here and there. But Rathi is not able to stand up, despite having had a round of injections the day before. Madhuram climbs up a neem tree in a daredevil way to lop off some leafy branches and then presents them to her. She devours them eagerly – at least she has not lost her appetite, so that is a good sign. While she is immobilised, the rest of the herd now make noisy submissions to get moving. Madhuram opens the gate and, after the camels have taken their fill at the communal water tank outside, he and the herd march towards the forest. They will only come back after sunset, spending the whole day foraging on a range of trees and mingling with wildlife.

Meanwhile, I go back to my desk to read the news and check my Twitter account. The Climate Summit is on in Glasgow. Not much information is filtering out from the official blue section where governments negotiate the final text of the COP26. But many non-governmental agencies (NGOs) agreed, and especially youth groups, have zeroed in on livestock as the main culprit for climate change. It is not surprising since over the last few years, there has been a chorus of voices – including those of scientists with respectable affiliations – proclaiming that the world would be better off without livestock.

'You can't be a meat-eating environmentalist. Everyone must go vegan for animals and the planet,' campaigns PETA.[2] And Pat Brown, the CEO of Impossible Foods, says it's ludicrous that anybody in the world should eat meat and maintains ending animal agriculture is the fastest and most powerful way to restrain climate change. 'Farm animals are the most destructive technology on earth and almost entirely responsible for the global collapse of biodiversity,' he states, as reported by the *New Statesman*.[3]

I sigh and shake my head in disbelief. How can my camels be deemed a 'technology', and the most destructive one on Earth to boot, when all they do is browse on thorny acacia trees and convert their leaves into milk that nourishes kids with deficiencies in their diets and helps cure diseases? This is an entirely natural process, while manufacturing artificial meat and dairy is an enormously complicated procedure. Does PETA know that if we all went vegan, camels would become virtually extinct globally, except for a few sad specimens in zoos? What an empty and boring world that would be!

However, I feel sympathy as well. After all, it is true that farm animals have been recast by many livestock scientists and industries as mechanical devices that convert feed into food, rather than as living beings with desires, individuality and a need for the stimulation of their senses. And the way many of them are kept is totally against nature: amassed in huge numbers, confined and crowded into tight spaces, stuffed with concentrate and feed additives, then culled after a preordained number of days. They are produced in an assembly-line approach, just like cars or furniture or any other factory-produced consumables. Huge amounts of grain that could be eaten by people are diverted to them. This feed/food is grown in monocultures, wiping out biodiversity while burning up fossil fuels that heat the atmosphere and using agrochemicals which poison the soil. The excrements of the living machines pollute air, water and soil, often making the environment around large livestock operations unlivable.

This is the current scenario, and it has rightly captured the attention of the media, the general public and especially the young. But it is only one side of the farm animal story. Largely out of sight is a parallel universe in which livestock are husbanded as a part of nature instead of being produced in factory-like surroundings. At the margins of our agrarian world, there is a myriad of herding cultures who regard farm animals as co-creatures rather than objects and whose relationship with them is based not on exploitation, but on reciprocity. They engage with livestock in ways which are simultaneously good for animals, people and the planet, and, in fact, they are essential to upholding the web of life on Earth and ensuring its future functioning. The Raika are an example, but they are just one among countless others.[4]

This book is about the hidden realm of peoples and cultures who herd animals, ranging from alpacas to yaks, and move with them through landscapes that are often remote and spectacular, but can also be intensely cultivated or even densely populated. And who, most importantly, look at their livestock as partners and treat them accordingly. Anthropologists refer to them as pastoralists, a term that has its roots in the Latin verb *pascere* which means 'to lead to pasture, set to grazing, cause to eat' and goes back to the Proto-Indo European root 'pa', meaning 'to protect, feed'. A pastoralist is defined as 'anybody whose livelihood comes from tending grazing animals'.

While industrial livestock production is less than a hundred years old, pastoralism has been around for close to 10,000 years.[5] A multitude of fascinating studies from disciplines that include ecology, range management, soil science, anthropology, development economics, environmental sciences and nutritional studies indicate that if we learn from pastoralists how to manage animals and the environment, we can address many of the problems currently facing humanity: the consequences and further worsening of climate change, the spiralling loss of biodiversity, and the public health issues that arise from misguided diets, antimicrobial resistance and newly emerging diseases. If we amend the human relationship with farm animals according to the pastoralist model, so that livestock emulates the role of wild herd animals in the landscape, there is no longer any rational reason for it to cause the ire of animal-rights activists or serve as a smokescreen for people and corporations who seek to make a fortune from artificial meat and milk.

Instead of worshipping 'good' wildlife and eliminating 'bad' livestock, we need to break down the artificial divide between them. They can coexist and, by skillful management, we can achieve their integration.[6] As is demonstrated by the livestock conservancies in East Africa, there are tensions and trade-offs, but it can be done. Furthermore, the effect of grazing by livestock is not noticeably different from that of wildlife, as long-term monitoring in Kenyan rangelands has shown.

---

The first part of this book introduces the herding universe: an array of cultures that live, and move, in joint households with their herds. It explains the importance of their holistic knowledge, especially in adjusting to current unpredictable times, and traces how a special bond developed between people and animals in prehistory that is not a given and needs to be continually refreshed. We look into how pastoralists magically transform waste into food and recyclable natural materials, why and how herders move, their significance for crop cultivation and soil fertility, and for sustaining diverse life on Earth. It draws mostly on my experience of living among the Raika for the last 30 years.

In the second section, I will look at how herding, by filling in for and mimicking the role of the herbivorous animals that once roamed the Earth in huge numbers, can help us repair the planet and deliver solutions to the environmental and other challenges that we face. It is based on my work with the League for Pastoral Peoples, a research and advocacy organisation that I initiated in 1992, originally to support the Raika but which then became engaged in international advocacy for pastoralists in general.

Finally, I present a perspective on livestock keeping that is fit for the future and will restore our relationship with farm animals to one of caring coexistence.

PART ONE

# The Herding Universe

# A Tapestry of Cultures

*The relationship between me and my flock is a spiritual relationship.*

ABU ALI, shepherd in Lebanon[1]

NOVEMBER 2010

I am in a taxi with four other women, driving on a bumpy, non-tarmacked road through a flat and rather monotonous stretch of agricultural land about a hundred kilometres north of the mega-city of Ahmedabad in Gujarat in India. The colourful finery of my travelling companions identifies them as Raika: wide swinging skirts printed in a traditional pattern, shiny red veils with a blue border enveloping tightly cut and embroidered blouses that leave their mid-riffs exposed. Their jewellery is even more eye-catching: wheel-like nose rings about 5cm (2in) in diameter, black neckbands from which dangle a row of small bronze filaments shaped like tiny feet, both upper and lower arms covered in white plastic rings, interspaced with rhinestone-covered bracelets at the elbows. We are on our way to a place called Mera where the first Global Gathering of Women Pastoralists is to take place. I am escorting the women as an interpreter.

It has been less than a decade since pastoralists made their first tentative steps towards establishing an identity as mobile indigenous people at the international level. It began with the attendance of Uncle Sayyad, a tribal elder from Iran at a major nature conservation meeting, the World Parks Congress in Durban, South Africa, in 2003.

In his speech to the international audience of wildlife enthusiasts he emphasised that, for his tribe, conservation was not a professional activity to remedy what had gone wrong, but an integral and essential part of their way of life.[2] Since then, there has been a series of irregular meetings, often at the sidelines of international conferences, where the respect of pastoralists for nature and the value of their traditional practices for the maintenance of the Earth's ecosystems and biodiversity were highlighted. Initially these meetings were organised by outsiders, but in 2007 the World Alliance of Mobile Indigenous Peoples (WAMIP) was founded in Segovia, Spain, to represent their interests.[3] These get-togethers have always been dominated by men with very few women in attendance. So, the gathering we are about to attend is a novelty; it is the brainchild of an NGO called Maldhari Rural Action Group (MARAG) that works for the rights of the Maldhari, which is the Gujarati term for pastoralists. Although they had the help of a couple of dedicated people at international agencies, it must have been challenging to pull off, I think, considering the problems of identifying women leaders in remote areas and organising the visa and travel logistics for around a hundred of them.

Among the women I am chaperoning is my friend Dailibai, whom I have known for almost 20 years. A midwife and animal healer, she never had the chance to become literate, but her enviable oratorical skills put most politicians to shame. She is an experienced community representative and has travelled the world to speak up about various issues around the Raika, such as their role in stewarding indigenous livestock breeds, the value of traditional knowledge and the need to conserve it, and the danger of corporate control over genetic resources. While she is a seasoned advocacy hand, the other three ladies, all in their forties and fifties, have never travelled anywhere beyond their native home and the village into which they married. It has taken Dailibai a lot of coaxing to convince them to join the gathering. They were keen to travel, but in Raika society women do not usually go anywhere without a male escort. In addition, there was the very practical problem of who would take care of their sheep- and goat-related chores during their absence: milking, collecting forage for the lambs, cleaning the pen. But curiosity finally won and here we are, approaching the venue on a narrow dirt road. From a distance we can see a throng of people crowding around the entrance gate, women from the local Rebari community who share many similarities with the Raika, welcoming and garlanding arrivals. Getting out of the car, we are engulfed in a riot of women dressed in diverse and colourful ethnic attires. In this melee, I recognise a few old friends, among them Elizabeth Katushabe, a Bahima from Uganda who is a passionate breeder of the famous long-horned Ankole cattle and a very articulate speaker about pastoralists' rights, as well as Khadija from the Iranian Centre for Sustainable Development and Environment (CENESTA), one of the pioneers in bringing pastoral issues to the international conservation stage, and Neeta, the driving force from MARAG. It is a joyful and emotional reunion totally unlike conventional conventions where everybody is wearing a business suit and which can be quite awkward if you don't know anybody. After hugging and being hugged by an Italian goat herd, a Mongolian representative and many other complete strangers, we are led to our accommodation in traditional round huts called *jhoopas*.

In the early evening, the opening ceremony takes place in an amphitheatre-like assembly hall that has been dug out from the earth. There is some simultaneous translation between Hindi / Gujarati and English through earphones, but the diversity of languages requires group-whisper translation. The organisers give speeches; Carlo Petrini, the founder of the Slow Food Movement, gives a motivational talk; and each country or ethnic group of women introduces themselves. In the almost exclusive company of women, my Raika companions have totally relaxed and take to the microphone like born entertainers.

In the night, there are unseasonal torrential rains that turn the amphitheatre into a pond and flood the round huts. Our luggage is drenched and, dripping with water, we are evacuated to makeshift arrangements in a nearby school. But this only adds to the ebullient mood and spirit of cooperation. Pastoralists are used to taking extreme weather events in their stride and this inconvenience does not distract them from the discussion of urgent matters which takes place in working groups organised according to geopolitical areas and themes. We hear how governments either ignore herders or are settling them, and how mobile livestock keeping is branded as backward and unproductive by policy makers, while scientists blame it for causing desertification and destroying the environment. Many of the participants describe how it has become almost impossible to continue in the old ways and emphasise that there is a need to explore new economic options. They universally praise the importance of retaining traditional wisdom and value systems, while pointing out that it is regrettably no longer passed on to the next generation.

At the end of the meeting, the women agree on an official outcome, the Mera Declaration.[4] It is quite an elaborate document, so here are some of the extracts:

> We, the women pastoralists gathered in Mera, India, from November 21–26, 2010, representing 31 countries, have met to strengthen alliances and forward practical solutions to issues that affect us.
>
> We are part of a world-wide community of pastoralist peoples that is 300 million strong. We pledge to continue to live in a way that is environmentally sustainable and protects biodiversity and common resources for generations to come.
>
> We experience firsthand the leading edge of climate change and its associated problems, and we have much to share with the world about adaptation, mitigation and living sustainably on planet earth…
>
> We women pastoralists want our children, and our children's children, to have the tools and opportunities they need to adapt to the realities and changing conditions of the modern world while retaining their traditional cultural legacies and lifestyles.

> This is our right and it is by remaining pastoralists that we can be of greatest service to the entire human community.

---

The women assembled in Mera came from 31 countries, ranging from Afghanistan to Uzbekistan in Asia, including many African nations, and a couple from Europe and the Americas. Herding cultures are spread all over the world, although they are dominant wherever land is not suitable for growing crops. Those of us hailing from countries with temperate climates and regular rainfall patterns rarely realise it but, in most of the world, conditions are too dry, too hot, too cold or too mountainous to practise sedentary agriculture.

Land suitable for the plough is concentrated in Europe and the eastern part of the United States as well as in the Indian subcontinent. A substantial patch is present in northeastern China, and Australia has cultivable edges in its southwestern and southeastern corners. In Africa, the Ethiopian Highlands stand out as the most prominent locality, with little blotches in South Africa, and there is a streak of fertile land in Argentina that extends inwards from Buenos Aires. Altogether cropland only extends over around 14 per cent of the ice-free land of the Earth, amounting to over 1.407 billion hectares.[5] This represents less than one-third of the 4.924 billion hectares that are classified as 'agricultural' land by the Food and Agriculture Organization of the United Nations (FAO). The term 'agricultural land' might conjure images of a fertile countryside and bountiful harvests. But this land category comprises not only arable land fit for growing crops, but also land under permanent crops (such as orchards) and 'permanent pastures'. The latter land type takes up two-thirds of the agricultural land and here the only option for producing food is by keeping livestock that upcycle the native vegetation – which may be grass, but can also be sparse, thorny and fibrous shrubs – into milk, meat, fibre and a range of other products. In partnership with animals, humanity can produce food in the frozen taiga and tundra, seasonally get by in the Arabian Desert and, at high altitudes in the Himalayas, even use both sweet- and salt-water bodies for generating protein. While growing crops is feasible only in a few privileged areas of the world, almost no place on Earth is out of range for livestock.

The type of land that is not cultivated but suitable for grazing and browsing by animals – both domestic and wild – is referred to as 'rangelands'. This category encompasses a wide array of ecological zones, such as deserts, savannahs, steppes, tundras, prairies, woodlands and shrublands. In these huge swathes of land that have not been repurposed for growing crops, nature still holds sway over people, and they include many of our most iconic landscapes, such as the Tibetan Plateau, the savannahs of East Africa, the steppes of Mongolia, the alpine meadows in Switzerland, Austria, Germany and Italy, the Carpathian Mountains in Romania, and the Lake District and the Yorkshire Dales in the U.K. These landscapes support 50 per cent of the world's livestock and while they might appear 'natural', they have been shaped, if not created, by herding. Although they personify nature, and are often teeming with wildlife, they are 'human-livestock'-shaped and cover more than half – 54 per cent – of the Earth's surface.[6] Described as a 'source of beauty and continuous fascination', they were long protected from destructive human manipulation, but mining and energy generation have turned into major threats to their existence.[7]

Yet pastoralism is not limited to rangelands; it extends much, much further. Herding plays an extremely significant ecological role in many crop-cultivating areas, as a means of utilising the aftermath and converting it into organic manure. And even forests are an important herding habitat in South Asia, as are woodland pastures in Europe.

So, thinking of pastoralism as a minor and marginal phenomenon is a view from the 'North', a figment of our biased urban perceptions. In reality, it extends over a far greater share of the globe than sedentary farming. Together, herding cultures manage at least four times as much land as sedentary farmers cultivate.

Nevertheless, sedentary farming is regarded as the default model for global food production, while pastoralism is relegated to a status of insignificance. There are several historical reasons for this.

An early factor was a deep-seated fear of 'marauding nomads'. Before the present era of almost universal surveillance, pastoralists could be here today and gone tomorrow; they had the capacity to move into remote areas to extract themselves from the clutches of central powers and stay incommunicado, so that nobody knew which possibly unlawful activities they were engaging in. There are many historic instances where mounted pastoralists used their mobility for regular attacks on settled communities and even to subdue them. Examples include the Xiongnu or Hsiung-nu, a tribal confederation composed mostly of pastoralists who ruled much of Central Asia in the second and third centuries BCE. They were a trigger for the sedentary Han Chinese to build the Great Wall of China.[8] In the thirteenth century, horse-riding Mongols overran Asia and much of the Middle East, creating the largest empire ever.[9] Camel-mounted nomadic Bedouin were long a force to be reckoned with in the Middle East, looting trade and religious caravans and challenging central control.

But it was in the colonial period that the trajectory for the institutional disdain and neglect of herding as a way of using land was set. During the period of European domination, the mental model of agriculture that originated in the rather small part of the world with year-round mild temperatures and regular rainfall was exported and imposed on much of the rest of the globe. The foundations were laid for many of the institutions that even today govern how pastoralism is viewed and handled. The colonial powers created government departments and research institutes that gave more importance to crops than to livestock. In addition, animal husbandry was held to be sedentary by default. These structures are still with us today, including in countries where animals are economically much more important than plants and where the majority of livestock is kept in pastoral nomadic systems. African countries such as Sudan, Ethiopia, Kenya, and many others, as well as India, are cases in point. While there are efforts in some disciplines, such as the geo-sciences, to 'de-colonise' them,[10] this has not happened in agriculture, and research and the teaching of agricultural sciences everywhere follow the 'Western model' in which yields are to be maximised, without any consideration of the negative fallout.

In order to illustrate the forces at work during colonial rule, let us take India as an example, for they have been documented beautifully by the historian Neeladri Bhattacharya.[11] He narrates that in pre-colonial times, pastoralism was a highly valued and indispensable component of rural life, and there was a reciprocal relationship between herders and farmers. The former bred bullocks and male camels that farmers needed to till their fields and for other agricultural chores. They also provided manure. Often there were exchanges in kind, such as manure being traded for a sack of grain. Designated grazing areas, in the form of village grazing grounds, sacred groves, reserve grazing areas or just open land, were part of the landscape and livestock cycled nutrients from these uncultivated areas to farmers' fields. Such mutualistic, synergetic systems prevailed throughout the sub-continent. When the colonisers, on a mission to control, tax and extract wealth, made their entry, they had a certain mental picture of what a landscape should look like. The vision of administrators was based on farmland in temperate and evergreen England. For them, all uncultivated land was 'ugly', 'waste', 'barren', a 'wilderness'. In their view, Nature was an enemy that had to be conquered. They held it to be their Christian duty to transform uncultivated areas by means of labour into orderly and fenced farms, preferably of wheat fields. Such transformation was a measure of progress and reflected the good deeds of civilising the native people and bringing order to their landscapes. Based on this worldview, it was mandatory to claim and cultivate the areas where pastoralists grazed their livestock.

One herding group that felt the transformative zeal of the British were the Gaddi shepherds in the Himalayas. They had traditionally adhered to a combination of private and communal rights that regulated access to grazing. Their elaborate system of managing grazing on fields, in forests and on alpine pastures included mechanisms for conflict resolution.

The colonial state declared all land that was not cultivated with crops to be state property. This included uncultivated areas that the British termed 'wastelands', as well as forests. From then on grazing fees had to be paid for the use of both land categories. Colonial administrators also interfered with the timing of herd movements, setting fixed dates for their arrival and departure at certain locations, without considering the annual fluctuations in weather which the Gaddis needed to take into account when moving their herds. If they did not adhere to the migration dates prescribed by the government, they were fined. Earlier, the Gaddi sheep flocks had provided manuring services to landowners, but suddenly the official calendar no longer provided time for this.

In the opinion of the colonisalists, pastoralists were lazy, lawless, wild, cowardly and roaming around without purpose. In 1871, a Criminal Tribes Act came to pass that declared wanderers to be criminals and mandated them to stay put in villages. Nobody could leave without a licence. The list of criminal tribes included many pastoralist groups.[12]

Yet, it was by no means just the British, or West European colonisers in general, who had this attitude. In the early twentieth century, the communist Russian state adopted a similar approach in the nomadically managed areas that it took control of. In Kazakhstan, a programme to sedentarise its largely nomadic-pastoral population and confiscate their livestock led to a great famine from 1931–1933 in which an estimated one to two million people – or about a third of the population – died, with several hundred thousand fleeing to Mongolia, Siberia, Uzbekistan, Iran, and other places.[13]

In Siberia, the Russian intention was to civilise the nomadic reindeer herders by building villages for them, providing services and teaching them communist values. All children from the age of seven were sent to boarding schools and selective youths were taken for training at a special college. The shamans who upheld the connection between the reindeer herders and the spirits of the land were killed.[14]

For China, nomads are a nemesis that they have been struggling with since at least the fifth century when they started building the first elements of what eventually grew into the Great Wall of China. Throughout its long history, sedentary China had hostile interactions with nomads, and a fear of them may have become deeply embedded in its psyche. More than 40 per cent of China's territory is covered by grasslands that are traditionally inhabited and managed by nomadic pastoralists. In the opinion of the government, pastoralists cause overgrazing, destroy the environment and interfere with wildlife conservation. They must move into permanent houses.[15] Nomad sedentarisation projects were important elements of its 11th and 12th Five Year Plans from 2006–2016, leading China to institute some of the most repressive regimes for pastoralists in current times. They have led to ecological damage and social deprivation in Sinkiang, Inner Mongolia and on the Tibetan Plateau.[16]

This skewed perception is prevalent in the minds of the people who dismiss the necessity of livestock for human existence, despite tens of millions – maybe more than a billion – of people in rangelands depending on them. The colonial legacy lives on in the outrageous statements by the CEO of Impossible Foods who aims at eliminating livestock from the face of the Earth within the next few decades, and it is also reflected in the papers by scientists from respected universities who paint rosy pictures of a future without livestock. An article written by Oxford University-based researchers and published in the journal *Science* claims in all seriousness that 'without meat and dairy consumption, global farmland use could be reduced by more than 75% – an area equivalent to the US, China, the European Union and Australia combined – and still feed the world.'[17]

The women gathered at Mera would have been both aghast and incredulous at such statements. So would the pastoralists all over the world who herd reindeer in the Arctic, manage yaks in Asia's high-altitude zones, keep Bactrian camels and dromedaries in the deserts of Asia and Africa, move around with cattle, sheep and goats in the semi-arid steppes and savannahs of Africa, and husband llamas and alpacas in the Andes in South America. Life without livestock is unimaginable to them on both a spiritual and material level. Their relationship with animals is the foundation of their identities and worldviews; livestock are their means of upholding community ties and helping others, as well as the basis of independence from bosses and overlords. For them, animals are often not just a source of food, but also of fibre for clothing, shelter and interior decoration; of fuel for cooking; a means of transportation; and, of course, income. Pastoralists care for, respect and even venerate their animals – as behoves a partner without whom there would be no life. This love and respect are not sentimental, nor reflective of the kind of concern that the animal-welfare community espouses, but is created by mutual dependence. People and animals are in it together: without the other neither can survive.

Here is how a pastoralist woman describes how she feels about the sweaty and back-breaking work of collecting fodder for her calves, a task that requires walking long distances over rough terrain, cutting a bundle of different plants and then carrying home the heavy load:

> The reason why I feel happy is: I don't beg. I just herd my animals and when they come back in the evening I just milk. If my family is hungry, I can slaughter for them. During the rainy season when there is good milk, I process to get butter from them. That is why I don't even feel the hardness of these jobs.[18]

The woman quoted belongs to the Borana, a cattle-herding group residing in Northern Kenya and Southern Ethiopia that is famous for a unique community governance system and for its very complex institutions around managing animals, grazing and water resources in an equitable way. Traditionally, the labour around livestock was shared between men and women, but with many men seeking outside work, a whole range of responsibilities now often falls on women.

Such alliances with livestock underlie human existences in most of the world and, to give you a sense of the enormous spread of partnerships across continents and cultures, here are some of the most intriguing examples:

**Reindeer.** In the circumpolar region, reindeer are the source of life for a total of 20 different ethnic groups, including the Sami, Evenki, Eveny and Chukchi. Ferreting out forage in the tundra and taiga, including the leaves of birch and willow trees, lichens and reindeer moss, even under a deep layer of snow, reindeer provide their herders with food (milk and meat), clothing and shoes (made from their fur), transportation (pulling sledges or as riding animals), housing (tents made from skins) in temperatures that can be as low as -71°C (-96°F). These antlered creatures do not just serve up material benefits, but also nourish the souls of their keepers and protect them from evil. British anthropologist Piers Vitebsky, who spent decades living among the Eveny in Siberia and Eastern Russia, reported that, in earlier days, the Eveny undertook annual soul-voyages to the Sun on the back of their reindeer, from which they returned tired but happy and relaxed. He hypothesised that it was the reindeer that had domesticated people, instead of vice versa.[19] Herded reindeer have a much longer lifespan than their wild relatives, on average living to the age of 18 years instead of ten.

**Yaks.** I encountered and rode these shaggy creatures during a short trek in the Indian Himalayas. I was surprised how comfortable they are to sit on and how easy to handle. Unfortunately, they only do well at higher altitudes of 3,000–5,000m (10,000–16,000ft) above sea level. In the high mountains of Asia, the Pamir and the Himalaya-Hindukush range and on the Tibetan Plateau, they are the foundation of the livelihoods of around thirty ethnic groups for whom they provide transportation, clothing, furniture and housing, fuel, and food.[20] The significance of yaks goes beyond the material: their Buddhist keepers regard them as safe-guarding gods and assign cultural and religious significance to each of the yak's body parts. Yak heads are placed on the walls of houses and sometimes a yak carcass is hung up to ward off evil spirits. Yak butter sculptures are burned as offerings to the gods and can be found in most of the monasteries of the Tibetan area.

**Water buffaloes.** I have encountered buffaloes almost everywhere in India in a range of environments, from the alpine meadows of the Himalayas to the arid expanse of the Kutch Desert, the lagoons of the Bay of Bengal to the jungles of the Deccan Plateau. They are a lifeline to both Hindu and Muslim herding communities. The most treasured product of these sluggish-looking animals is milk with a very high fat content. I have had the pleasure of meeting the Van Gujjars, a community of Muslim buffalo breeders who seasonally migrate from the foothills to the alpine pastures of the Himalayas to ensure the well-being of their herds. They treat their animals, which also include ponies to carry their belongings, as family members. American writer Michael Benanav documented how the young men of a Van Gujjar family carried a young buffalo with a broken leg over long distances on their shoulders so they would not have to abandon it.[21] Indian veterinarian Balaram Sahu described how the buffaloes of the Chilika Lake in India swim out to graze on shifting islands where their presence attracts crowds of migratory birds who feed on the insects that multiply in the dung and the puddles left by hoofprints.

**One-humped camels.** My favourite animal, the one-humped camel, is at the centre of pastoralist societies in the harsh deserts and steppes that stretch out between Mauretania on the west coast of Africa to the Thar Desert in India. The largest concentration of one-humped camels is in the Horn of Africa where camel milk is the dietary staple of the Somali community. A common practice among these diverse groups was never to sell camel milk, and only give it away for free to people in need. The Afar from the drylands of Ethiopia put it like this: 'If a son dies, he is buried, but life goes on and you may have another son. But if a camel dies, everything comes to a halt and without movement there is no life and the household breaks down.'[22]

**Two-humped camels.** In the Gobi and Taklamakan Deserts in Mongolia and China, the two-humped Bactrian camel thrives on salty, dry, thorny and bitter plants. Turkic and Mongol nomads process camel milk into a range of dairy products, including cheeses and vodka home-distilled from fermented milk. The long, fine hair of the Bactrian camel is a much sought-after raw material for making luxury garments.[23]

**Cattle.** Practised in many parts of the world, pastoralism is – or was – especially prominent in India and Africa and has also given rise to many cattle cultures. Among them are the Fulani, a group of cattle herders who inhabit the Sahelian belt that stretches across Africa from west to east. One of the Fulani subgroups, the WoDaaBe of Niger, have bred the Bororo breed of cattle which amazed colonial veterinarians with its ability to thrive during a nine-month dry period and on pastures that seemed unproductive. The breed has survived major drought periods and is highly desired for export, according to Saverio Krätli, an Italian-born social scientist who has studied the relationship between the WoDaaBe and the Bororo cattle in depth.[24] Cattle pastoralism is also practised in Spain where hundreds of cows trek 300km (186 miles) from winter pastures in the south to summer grazing in the mountainous interior in a journey that takes 15 days. In Northern Greece, especially in Thessaly, cattle traditionally stayed in the lowlands during the winter and moved into the highlands in the summer. These systems were focused on meat production, required no capital and absorbed family labour.[25]

**Sheep.** Found all over the world, sheep are kept by Muslims, Hindus, Buddhists and Christians over a vast range of ecological zones that range from extreme drylands to places with exceptionally high rainfall. The Bedouins of the Syrian Desert bred the Awassi sheep, which stores fat in its tail to see it through prolonged droughts. At the other end of the rainfall spectrum, the Kuruba pastoralists of the Western Ghats in India have developed the black Deccani sheep, a breed that is impervious to the extreme wetness resulting from annual rainfall amounts of 900cm (350in) pouring down during the few months of the monsoon. The wool of this animal is woven into ritual blankets that are necessary for the life cycle rituals of the Kuruba. Community members take care of a huge flock of sacred sheep. The same group of shepherds worships wolves because they think they are essential for keeping their flocks healthy by culling weak and sick sheep.[26] Seasonal sheep movements are widespread in almost all countries of Europe, notably in Spain and in the Carpathians.

**Horses.** An animal of the Eurasian steppes where they were first domesticated, horses have gradually been adapted to totally different environments, ranging from the icy climes of Siberia and Iceland to the torrid heat of the Arabian Desert. Horse nomadism is a key feature of Central Asian culture, enabling the rise of the Mongol Empire. In Mongolia, horses are not just a source of transportation but also of food. Fermented mare's milk is popular among many Central Asian societies. In Europe, there are several instances of horses kept in free-ranging systems in which they are rounded up once or twice annually, such as in the Camargue in France and in Iceland. The Mérens horses of Southern France were bred in transhumant systems, moving to summer pastures in the Pyrenees during spring and returning to the valleys in autumn. The practice is being revived and attracts tourists who accompany the trek.

**Donkeys.** Originating in Northern Africa, donkeys first came to prominence in ancient Egypt. While few herding societies centre on them, they are often used as docile adjuncts for carrying household belongings during migration, including by the Raika in Rajasthan and the Turkana in Kenya. In the Sicilian town of Troina, the revival of the ancient practice of moving donkeys from the plain to uplands during the summer months enabled citizens to wrest control over forestland from the Mafia and to provide employment for young people, while also conserving the Ragusano donkey breed and the Sanfratellano horse.[27]

**Llamas and alpacas.** In the South American Andes, llamas and alpacas were domesticated by the Incas, diminished by the Spanish colonists, and are now kept by both Aymara- and Quechua-speaking pastoralists for whom the relationship with these animals is part of their identity. Alpacas were mostly kept for their wool and historically yarn was an important trade item, while llamas were used for transportation. The two closely related species were important for rituals while their meat was also valued. Alpaca herders in the South American Andes say, 'In the same way as we nurture alpacas, they nurture us' and 'the day alpacas disappear, the world will disappear.'[28]

**Goats.** Everybody loves curious goats, and they are easily the most adaptable and easy to manage animal. They form the basis of the culture and livelihoods of a diverse range of pastoralists on all continents. I have visited both the Bakkarwal pastoralists of Kashmir in India[29] and the Crianceros of the Argentinian Andes, both of whom cover large distances between specific summer and winter grazing sites, a practice referred to as 'transhumance' that is also followed in many European countries, including Greece, Spain, Switzerland and Austria, usually in combination with sheep. But goats are also making inroads into conservation grazing to control weeds or prevent forest fires, and they are widely used for this purpose in California where they were recently reported to have saved the Presidential Library of Ronald Reagan from burning down.[30]

---

Because pastoralism comes in such a diversity of guises, it evades a succinct definition. Attempting to cover pastoralism's many permutations, degree and pattern of movement, extent of integration with crop cultivation, and as a source of livelihood, expert definitions can become very long-winded. Some of the simpler ones are 'A pastoralist is anybody whose subsistence depends directly or indirectly on the grazing of animals on native pasture' and 'Pastoralism is a social and economic system based on the raising and herding of livestock'.[31]

But the most elegant and pertinent definition I have come across was ventured by a Fulani herder during a pastoralist meeting in Burkina Faso, in West Africa, that I attended many years ago: 'Pastoralists are people who have a social relationship with animals.'

This definition goes to the core of the issues discussed here. Pastoralism, as understood in this book, is centred on the personal relationships between people and animals. It entails family friendships between human and non-human animals that usually go back over generations – maybe even in an unbroken chain to the mythical event in which God created the human ancestor for the purpose of taking care of livestock. Pastoralism means living together in the same household; it involves intimacy through daily interaction. To make pastoralism work, herders must basically subjugate their own comfort to that of their animals. The herd always comes first. German master shepherd Ruth Häckh, who herds her family's sheep flock on the Swabian Alb, described it like this:

> When I trained as a shepherd, we were taught: the flock comes first, it comes second, and it comes third. Then come the sheepdogs, and then for a long time there is nothing, and finally there is you.[32]

Animals kept in herding systems benefit from the relationship just as much as people. They live longer than their wild relatives and usually (although not always) much, much longer than their kin in intensive and industrial livestock systems. Their keepers protect them from predators, from insects, from the elements. They ward off evil spirits and court benign ones. They nurse them when they are sick. They bottle-feed baby animals that have lost their mothers. They rescue them when they are caught on a cliff, fall into a well or get lost in the forest. They help them access forage by scouting out vegetation, lopping trees for fodder or removing ice layers. They lift water for them from deep wells with immense physical toil. In the difficult terrains that are their homelands, animals would not exist without their herders, nor people without their livestock. Both parties share the benefits. The usual arguments that the animal rights movement brings forward against animal farming as a system of exploitation do not apply to pastoralism.

This does not mean that pastoralists are perfect or never do wrong, or that their ways cannot be improved. For one, as in every profession, there are some who are extremely skilled and hard-working, and others who are less so. Secondly, herding cultures have rapidly unravelled in many places and are now only a shadow of their former selves or have even disappeared entirely. But the ethics and practices of working with nature that have been the foundation of herding cultures for millennia form an antithesis to the dominant concept of land as an asset from which resources are to be extracted and of animals as machines from whom profits are to be maximised. That is the reason why pastoralism is so important, although it is not the only one by far.

One more point. The fact that the term 'pastoralist' cannot be clearly defined can cause confusion. Ranchers in some countries such as Australia and elsewhere also call themselves pastoralists. The big difference between ranchers and the pastoralists that feature in this book is that the former own the land on which their herds roam while the latter have no titles to land, only customary rights (which is the reason why they are vulnerable). They depend on what is termed 'Commons', meaning land that is not privately owned, but to which groups of people have had customary rights for generations or 'time immemorial' based on local norms and agreements.

I have nothing against ranchers, as they share certain features with the pastoralists here, such as keeping animals moving and fed on natural vegetation. But there are also important differences and I want to focus on the unique aspects of indigenous herding communities, on their systemic understanding of how the different components of life on Earth interact and their knowledge of how best to arrange oneself with nature.[33]

# Knowledge of the Whole

> *Pastoralism is an intelligent production system with no equal. It takes uncertainty as an input to clothe & feed humans and nurture agricultural fields.*
>
> GOPI KRISHNA, Mitan[1]

So, what is so special about pastoralism? What makes up the cultural legacy the women refer to in the Mera Declaration and that they worry will be lost? And why do they say that by remaining pastoralists they can be of greatest service to the entire human community?

A big part of the cultural legacy – and the foundation of herding cultures of the world, whether they are reindeer herders, migratory shepherds or camel nomads – is an extensive body of knowledge of, and expertise in, managing livestock in tune with the environment and as an integral part of the landscape and local ecosystem.[2] This is paired with an ethical framework that conceives humans as a part of nature, rather than apart from and in antagonism with it. In their worldview, humans, livestock and wild animals are almost on the same level, and not arranged into a hierarchical order in which people reign supreme.

I will try to describe this by using Madhuram and the Raika community as an example, because I know them best and am most familiar with their ways. I want to add two caveats, though: firstly, I do not claim to do justice to Madhuram's knowledge – that would be preposterous and much of it is intangible and cannot be translated into words anyway. Nor do I want to give the impression that every

Raika knows as much as Madhuram or shares his worldview. As most Raika have given up herding and the transfer of the community's traditional knowledge from one generation to the next has come to a virtual standstill, Madhuram is unfortunately the exception rather than the rule. But at least we can get an inkling of what once was, as well as a sense of why it is important to revive and resurrect herding knowledge if we want livestock futures that are attuned to what our planet provides, do no damage to the environment and keep livestock in a way that suits their innate behavioural needs.

As a Raika, from a family well known for the quality of its livestock, including cattle, Madhuram believes that it is his inherited duty to take care of animals. The health and well-being of the animals in his care reflect his own skills as a herder. If they are in a bad condition, it means he is not doing his job properly, and it's a cause of extreme embarrassment that affects his and his family's standing in the community. If you ask him about his job, he will reply '*unto ki seva karna*', which means 'to serve the camels'. In the last few decades, as more and more Raika have looked for alternative employment, much of the community's livestock-related mores have faded away, but not so long ago, they adhered to an internal law not to build proper houses because this would interfere with the mobility and therefore the well-being of their herds. As was the case with selling female animals, transgression against this rule was punished with what is known as social boycott or outcasting, meaning exclusion of the family from all social interaction with the rest of the community by edict of the Raika elders. (The times when these actions were punishable are gone, and now every Raika aspires to build a big concrete structure that is in line with local ideas about what a house has to look like. The elders, however, still impose social boycotts on any family whose offspring marries out of caste.)

But 'serving the camels' is still the basic moral framework of Madhuram and his herding peers. Embedded in this is Madhuram's knowledge of animal behaviour and the nutritional value of the trees, shrubs, flowers, vines and grasses on which camels forage, as well as how these change during the seasons and affect milk yield and quality. He can predict weather patterns based on the behaviour of birds and other wild animals. He discerns subtle changes in the environment, such as in the composition of the local flora that nobody else will notice.

Students of academic animal science learn how to calculate feed rations to ensure that livestock receives just the right amount of protein, fat, minerals and other essential nutrients, and all this at the lowest market rates. The Raika have no concept of this, but they do have very clear ideas about what constitutes a good diet for their animals and immense botanical knowledge of the forage plants in their respective areas. This is combined with an understanding of how these plants influence milk quantity and taste, and whether they have other effects, such as causing bloating, or contain sufficient salt content. Herders can judge the quality of the forage from the status and behaviour of their animals; they do not need to send a sample to a lab for chemical analysis to determine whether some vital ingredient is missing.

Among the items on the camel menu are such diverse offerings as the pods of thorny acacia trees, thistles growing on fallow fields, vines that flourish during the monsoon, leaves of the neem tree and coarse saltbushes, to name just a small selection.[3] Like in a good eco-conscious organic restaurant, the menu is seasonal. In the area where Madhuram operates, camel diets are determined not just by rain, but also by local crop cycles. Social arrangements with farmers are crucial for Madhuram to ensure good nutrition for his camels throughout the year on a mosaic of different types of land. For Madhuram owns no land himself and depends on access to what is known as 'common pool resources': harvested fields, forest, land that belongs to a temple, village grazing grounds and what, in a relic from colonial times, is officially classified as 'wasteland'. In reality, managing a herd on disconnected pieces of land is like playing hopscotch with 40 large animals in tow that listen to you, follow you and trust you. It is no small feat.

Madhuram is an expert in tracking camels, a skill that is very necessary as it often happens that an individual, or a small group, is tempted to explore for forage away from the main group and disappears into the landscape. Some trackers are able to determine from footprints whether a camel is pregnant and how it is related to other camels. Madhuram's skills are not on that level, but are good enough for his purposes.

Madhuram knows the genealogy of each camel in his own herd and their degree of kinship with those in other herds in the vicinity. Like most pastoralists, the Raika conceive their herds as being composed of female bloodlines – matrilineages – whose foremothers have been a part of the herd for generations. All members of a line have the same 'family name'; to distinguish between them, they may have individual nicknames that reference a particular characteristic. The pedigrees of the camels are not written down anywhere, but Madhuram has them all in his head. Each lineage is known for specific traits – the Mewari family has good milk yields, the Rathi ladies are tough with good height, the Dholi lineage has good milkers and a light colour.

And Madhuram is also knowledgeable in the diagnosis and treatment of camel diseases, as are many of the old Raika. Herders are obsessed with the health of their camels and have, or had, an extensive disease classification system and a pharmacopoeia of traditional treatments in their head. They even used to do simple vaccinations against viral diseases such as pox, infecting animals with small doses at the time of year when the disease takes a more benign course. One of the most dangerous diseases is trypanosomiasis, an illness that is caused by a blood parasite and resembles malaria in humans. It often leads to miscarriages, even death. Madhuram uses the sandball test to diagnose it. This method involves waiting until the camel urinates and then forming the soiled earth into an orange-sized ball. After the ball has dried, it is broken up and from the smell it is possible to make a diagnosis. It is a method that amazed British colonial doctors more than hundred years ago. They pronounced it as more reliable than testing a blood sample under the microscope and referred to the Raika as native camel doctors.[4]

When Rathi, the camel that could not get up, did not respond to conventional treatment, a group of elderly Raika turned up to examine her and decided they needed to administer *dam*, which is the application of a hot iron to certain body parts. I had documented this practice some years ago and commented that the *dam* scars were like a patient file, indicating the diseases a camel had gone through in its life.[5] This led to an annoyed and irate response by the veterinary establishment who accused me of supporting cruel backward methods. And it is true, applying a burning hot iron is a painful procedure, but it also seems to work for some unexplained reason in a good number of cases. And if it saves a camel's life, and no alternative is available, it certainly has its merits. As it is a treatment that appears to be applied by many camel pastoralist communities throughout their entire range, it is certainly worthy of scientific exploration.

In any case, when I learned that Rathi was going to be treated with the hot iron, I discretely disappeared into the background, not wishing to witness her pain, but also not wanting to interfere with the local belief system. There was no alternative anyway. However, within minutes of having undergone the procedure, Rathi stood up and started browsing; she joined the herd on its daily round and has been fine ever since!

The entire herding knowledge is, or was, passed on orally from one generation to the next. Madhuram has learned from his father about camel diseases and their treatment, but it is unlikely that this body of knowledge will be transmitted further, since his son shows no inclination towards herding.

This knowledge is independent of all the virtual tools, apps and electronic gadgets that many of us have come to rely on to make decisions in our daily lives and that act as an extension of our memory. The fact that the older, herding generation of the Raika has no access to the written, let alone the virtual world, has enormous disadvantages, of course, as it makes it difficult for them to understand the larger political developments that impact their way of life and the economic contexts in which they operate. On the other hand, their knowledge is incredibly useful when responding in real time to any crises brought on by the natural world, possibly more useful than the masses of data, or even Big Data, we have accumulated in our virtual clouds and which we analyse to arrive at data-based decisions and to model developments and predict future events.

In our current age of specialisation, or even hyperspecialisation, the holistic nature of herding knowledge – which is really system knowledge – is precious. When I studied in veterinary school, information was doled out piecemeal by about thirty different disciplines, ranging from anatomy to zoology, and included physiology, animal nutrition, animal breeding, agricultural economics, microbiology, biochemistry, pharmacology, virology, parasitology, dairy hygiene, meat inspection and diseases of specific types of livestock, as well as gynaecology, obstetrics, andrology, and more.

Undoubtedly, there is a need to structure scientific disciplines into subcategories to allow for a greater depth of understanding. But many scientists have specialised to such a degree that they have lost sight of the overall picture and no longer understand the implications of their work for the system, or systems, as a whole. In the animal and veterinary science sectors, the animal breeding expert dare not utter an opinion on animal nutrition and vice versa. The parasitologist will have only a rudimentary knowledge of viruses and the virologist no clue about animal breeding. When I was studying, we never understood the overall context in which our lessons were embedded. We did not learn how to keep animals healthy, only to diagnose and treat their diseases. We did not consider the role of livestock in the environment, animal behaviour was not part of the curriculum, and we never gained an understanding of the whole farming system and the larger forces at work. Animal welfare was not an issue we dwelled upon.

Herding knowledge is founded on an understanding of how ecological systems – soils, plants, animals, people and the weather – articulate and influence each other. If we wanted to systemise it, we could also subdivide it into livestock management and nutrition, production, breeding and medicine, but it always comes in a package, as all parts are necessary to manage a herd successfully. It is based on continuous observations, entails flexibility to respond to changing circumstances and crises in real time, and is embedded in a value system that sees livestock as co-creatures and partners.

Madhuram's whole life is geared towards keeping his herd happy and healthy, doing everything in his power to prevent his wards from becoming sick. This care and concern for animals is missing from animal science, the academic discipline that is the intellectual foundation of modern, scientific animal production. Academic animal science conceives livestock as input-output machines whose efficiency is to be maximised to produce more food with less feed. Following this approach, livestock is decontextualised from its social and ecological environment, and no heed is paid to the other aspects of the system in which it is embedded, such as biological diversity, soil health, water purity and the socio-economic conditions of livestock keepers. An exclusive focus on yields and 'efficiency' has led to the inordinate growth of industrial livestock systems and the huge public backlash against livestock that was evident at the Climate Summit in Glasgow.

There seems to be no underlying value system in animal science, the academic discipline, nor is there much of that in its sister discipline, veterinary medicine. Animal science is oriented at making 'more from less' and up to now neglects to recognise farm animals as sentient beings. (If there are exceptions to this somewhere in the world, I would be happy to learn about them and revise my opinion.) During my veterinary training, we had to do a one-week farm practical on the college's teaching farm where our main activity was to remove dead birds from their cages. I learned to castrate piglets without any anaesthesia and to artificially inseminate cows who only got to walk once in their adult life – on their way to the abattoir. And it all seemed quite normal – part of the trade and becoming a member of the veterinary club. If you flinched or expressed distaste, you were considered weak and not up to being a veterinarian. Looking back, it almost feels as if systematic desensitization to animal suffering was at work.

Nor do animal scientists tend to engage with any of the externalities caused by the efficient systems they espouse. The profession prides itself on creating greater output (eggs, meat, milk) with fewer animals, but does not care to see at what cost this has been achieved: a loss of biodiversity, terrible animal welfare and the disappearance of rural livelihoods.

And because in animal science feed conversion rate and output are the standards against which everything is measured and the worth of a livestock system evaluated, herding is widely regarded as unproductive, while intensive and industrial livestock keeping is the aspiration. In the Western world, we glorify 'science-based approaches' for managing animals, the environment and crises of all sorts, including disease outbreaks. Scientific validity and scientific standing are conventionally measured by publications in peer-reviewed journals. Knowledge must be written down and data-based, and preferably be quantitative, measurable and based on repeatable experiments. Unless an issue can be 'metricised', it is not eligible for scientific purview. Articles in the natural sciences usually need to be based on an experimental setup in which hypotheses are tested; they are generically structured into the same sections which include introduction, materials and methods, results, discussion and conclusions.

The orally transmitted corpus of knowledge of pastoralists and other indigenous people cannot be squeezed into this standard pattern. Yes, an outsider can come and interview a specified number of 'knowledge holders', using a questionnaire, group discussions or semi-structured interviews, as anthropologists do, and then write a paper saying that informants knew x number of forage plants or y types of diseases. I have done this myself, but what I gained was a knowledge of the existence and depth of herders' knowledge, not the knowledge itself, and my disastrous record as a practical camel herder bears testimony to that (see *Communication*, page 57). That is why just documenting traditional knowledge does not really preserve it – it needs to be lived, applied, revised and adapted.

I would argue that by merit of its holistic nature, the knowledge of Madhuram may be more valuable for the sustainable management of the world's 'resources' than the typically highly specialised science that focuses on eclectic aspects of isolated phenomena or on modelling future events. Modelling is fine and useful, provided all factors that can determine future developments are considered, but if certain parameters are omitted or unknown, the results can be far off. And there are often factors that are unknown, as we have seen during the COVID-19 pandemic.

**Table 1.** Herding knowledge versus animal science

| | Herding Knowledge | Animal Science |
|---|---|---|
| *Focus* | The whole herd, from a long-term intergenerational perspective. | Individual animal and its performance during its lifespan. |
| *Context* | Animals are managed as part of the environment, taking advantage of its variability. | Animals seen in isolation from natural and social environment and kept in a controlled and uniform environment. |
| *Goal* | Keeping animal numbers in balance with availability of forage, never exceeding it for long. | Maximising individual performance without concern for where and how the feed is produced. |
| *Method* | Letting animals walk to their feed, so they can compose their own seasonal menus based on local resources. | Maximising performance and efficiency by immobilising animals and feeding them with calculated and standardised rations. |
| *Value System* | Respect for all life and regard for animals as partners. | Maximising profitability and, more recently, minimising the emission of GHGs (greenhouse gases) per unit of product. |
| *Effect* | Livestock population in tune with natural resource availability and planetary boundaries, mimicking ecological role of wild herbivores. | Exponential growth of the global livestock population, at the expense of tropical forests and natural grasslands. |

Very regrettably, too, science is often not impartial. It is frequently marshalled by interest groups to validate the outcomes desired by them, and in animal science, much research is funded by companies whose earnings depend on upholding industrial practices. This was evident in 2021 during the UN Food Systems Summit in which livestock was an extremely controversial topic. The issue was so contentious that the group in charge of putting together recommendations, the 'Sustainable Livestock Cluster', never managed to agree upon an official statement. The process was held hostage by industrial interest groups who referred to their 'science-based approach' to prove how sustainable their practices and production were and how they had achieved a significant reduction in the emissions of greenhouse gases per unit of livestock product. The effort was derailed when Philip Lymbery from Compassion in Animal Farming, who espouse animal welfare criteria, came in and refused to agree with these claims.[6]

---

For herding cultures, animals are at the centre of their universe. Their minds are absorbed with the well-being of animals and the rhythm of their daily activities is determined by ensuring that the herd is happy and healthy. If the herd is happy, they are happy.

Animals are at the centre of their social order and a measure of wealth and prestige, as well as a source of pride. The status of the herd – its health and beauty – reflects on their own status in the community. Animals are a kind of 'currency' in all social transactions, and therefore the composition of the herd is a reflection of, and determined by, their own social network. Most or all rituals and life cycle events are associated with animals and can't be performed without them.

When considering these pillars of the pastoralist worldview, we can see how they challenge the common view of farm animals as being pitiable objects of exploitation that need to be liberated. In pastoral systems, animals are partners who must be cossetted and even obeyed. But what are the roots of these deep bonds between pastoralists and livestock? This is the next question we will address.

# Bonding

*The Bedouins never let the foal drop to the ground at the moment of its birth, but they receive it in their arms and handle it with the utmost care for several hours; they wash it and stretch its delicate limbs and caress it all over like a child.*

JOHANN BURCKHARDT[1]

Farming animals and the process of taking their progenitors from the wild and changing them into livestock is commonly perceived and projected as an act of subjugation and exploitation in which humans outwitted animals and subsequently enslaved them. But again, this is a particularly Western perspective, and one that is not necessarily shared by other cultures. Many of the herding cultures of the world ascribe their association with livestock to divine acts in which God, or one of their gods, entrusted them with the responsibility of taking care of particular species of animals. Or they believe that animals themselves sought out humans and made the decision to stay with us.

Native North Americans had a relationship with animals that was based on balance and reciprocity rather than domination. The idea that they owned the animals never occurred to them. Because they believed animals were imbued with powerful spirits, it was unconceivable for them to treat them as commodities that existed merely for their consumption.[2] The *Diné* (Navajo Native Americans), who are at home in a spectacular landscape between their four sacred mountains in the Southwestern United States, are a fairly recent addition to the kaleidoscope of pastoralist cultures. Before the Spanish Conquest in the sixteenth century, the *Diné* were one among many native American groups that farmed maize and hunted bighorn sheep and other animals. Like many hunting cultures, they had a close spiritual relationship with the wild sheep, whose shed wool they also collected. But they re-invented themselves as pastoralists after coming into contact with sheep introduced to the area in the sixteenth century by Spanish conquistadors. While other tribes, such as the Apache, hunted the sheep for meat, the *Diné* started breeding them and they soon became the hub around which their spiritual and social life revolved. The *Diné* are a matri-centred society and most sheep were owned by women who obtained wealth and prestige by weaving blankets from the wool. Astonishingly, they erased all their pre-sheep life from their memories and associate the origin of their society with the local creation of sheep. In their creation myth, the fate of their community is closely intertwined with that of sheep. In one version of the Blessing Way, the *Diné* spiritual framework, Changing Woman, the wife of the Sun, created the sheep and the horse, and only four years later, when she became lonely, decided to breathe life into the Earth Surface people.[3]

In India, the Raika believe they were made by Lord Shiva to look after the camel that his consort Parvati had shaped out of clay. The Kuruba shepherds of Southern India tell the story of why their sheep are black: Parvati had hidden some of the sheep in an anthill because they were making trouble. The anthill accidentally caught fire and the sheep inside were singed. Subsequently, the Kuruba had to take care of them.

---

The implicit assumption that farm animals are creatures to be mastered and controlled has also coloured the theories of the scientists seeking to understand the process of domestication – the conversion from 'wild' to 'domestic' animal that happened in prehistory as long as 10,000 years ago. Just like 'pastoralism', 'domestication' is another one of those terms that has been notoriously difficult to define because of its shapeshifting end results. But its key feature is assumed to be human domination through a process that began with capturing and taming animals and ended with controlling their breeding. Juliet Clutton-Brock, one of the foremost scholars of domestication, defined it as:

> ...the keeping of animals in captivity by a human community that maintains total control over their breeding, organization of territory and food supply.[4]

Even more influenced by Western notions of animal husbandry, Hungarian domestication expert Sandor Bökönyi stated that it involves:

> ...the capture and taming by 'man' of animals of a species with particular behavioural characteristics, their removal from their natural living area and breeding community, and their maintenance under controlled breeding conditions for profit.[5]

Bökönyi distinguished between animal keeping and animal breeding, which meant 'purposeful selective breeding and the control of both quantity and quality of feeding'.

The assumption that domestication was a unilateral process in which humans took the initiative and had all the agency is increasingly being questioned, and some scholars now wonder who domesticated whom. For goats that turned farm animals some 10,000 years ago in Iran's Zagros Mountains and the Levant, Hans-Peter Uerpmann, a German archaeologist who studied domestication processes in the Near East, suggests that they may have domesticated themselves.[6] From my own work on goat domestication at 'Ain Ghazaal in Jordan, it seemed as if the process proceeded almost automatically after people had become sedentary. With respect to reindeer, Cambridge University anthropologist Piers Vitebsky muses in a similar vein, since reindeer under human care live much longer than their wild counterparts.[7] And Saverio Krätli, who wrote his PhD thesis on the cattle management of the WoDaaBe cattle pastoralists in West Africa, found that they do not exert control over their fierce cattle, which are much feared by outsiders. Rather, they interact with them in a persuasive manner, without any force. He entitled his thesis: 'Cows Who Choose Domestication'.[8]

After I realised that veterinary practice was not for me and I needed a new purpose in life, I worked as an archaeozoologist in Jordan, a job for which the anatomical knowledge gained during my veterinary education came in useful. Volunteering with various expeditions, my task was to identify and interpret the animal bone fragments that the archaeologists dug up in large numbers whenever they excavated a site. My first assignment was with an American college that worked in Pella, a large site in the Jordan Valley which was occupied for around 7,000 years. While the rest of the team was out digging in the dusty trenches, I was stationed in the excavation headquarters at a long trestle table covered with small heaps of bone fragments, each batch sitting on a tag that denoted the locus from which it came. I would identify and record the fragments of each pile by body part and by species, also noting whether the bone came from an adult or immature animal. Compiling this information systematically and tabulating it made it possible to deduce in what proportion the inhabitants of the site ate mutton versus beef versus pork versus hunted game, such as gazelles or hare, and how this changed during the different occupation phases of the settlement. From the age profile of the animals, it was theoretically possible to reconstruct animal husbandry practices and infer whether animals were used for meat, milk or wool.

The most intriguing archaeological project I was involved in was the unearthing of 'Ain Ghazal, one of the world's largest settlements from the Neolithic period or New Stone Age, which is located on the outskirts of Jordan's capital, Amman. 'Ain Ghazal, which means

'Spring of the Gazelle', dates from the late ninth to sixth millennium BCE and tells of the early history of farming, allowing us to trace the gradual transition from hunting to animal husbandry.[9] Charred seeds found there testified that the 'Ain Ghazalians cultivated emmer and einkorn, which were early forms of wheat. The animal remains were dominated by goat bones. In the earliest layers, they were mixed with the vestiges of animals that were clearly hunted or trapped, such as foxes and even badgers and pine martens. This fauna suggested a wooded environment in the vicinity. In addition, there was also a good percentage of bones from gazelles, which are associated with a steppe habitat. Notably, the proportion of wildlife progressively decreased during the occupation of the site and in the latest layers they had virtually disappeared, being replaced by domesticated goats, sheep, cattle and pigs.

In parallel with the bone profile, the culture also changed. Initially, people excelled in artwork and lived in well-built and nicely decorated houses with plastered floors. They fashioned spectacular human statues from plaster with expressive faces that are billed as the earliest examples of human sculptures and have now found a home in Jordan's National Museum – these can also be admired in the British Museum and the Louvre museums in Paris and Abu Dhabi. More than 9,000 years ago, these people lovingly decorated the skulls of deceased family members with plaster and bitumen, so they could display them in their homes – just as we put framed photographs of loved ones on a mantelpiece. But this culturally highly refined way of life did not last forever. By the time people had stopped hunting and advanced to a total reliance on domestic animals, the material culture had become impoverished, the figurines disappeared, and the buildings were poorly made shacks. Both culture and biodiversity had deteriorated and eventually the town was abandoned.

We gradually pieced this scenario together, continuously adapting theories and interpretations as new finds – in the form of architectural features, stone tools, human burials, grinding stones, clay figurines, charred seeds, pollen profiles and animal bones – were unearthed. Every afternoon, after eight hot and sweaty hours in their excavation squares, the dust-covered and dehydrated diggers came to my bone lab to drop off plastic bags of faunal remains on my table for me to clean, sort and identify. This was the time for brief exchanges about anything exciting that they might have dug up in their squares or interesting items I had found in their bone bags, maybe the tooth of a hyena or the pathologically deformed ankle bone of a goat. But what people most eagerly wanted to know from me was whether the goats that formed a major part of the diet had been hunted and wild or already domesticated and, if the latter, when did this happen?

Looking at the ends of the limb bones I had arranged by anatomical element in neat rows on my table, this was indeed a question that I kept turning over and over in my head. The earliest layers of 'Ain Ghazal were dated at around 8200 BCE and this was the general timeframe in which goats were believed to have been domesticated. Claims for herded goats had been made for roughly contemporaneous sites in Southwest Iran and nearby Jericho.

But what do we mean if we say an animal is domesticated? And how does it happen? Is it a long-drawn-out process that takes many generations and hundreds of years, or does it proceed quickly? In the archaeological record, a status of domestication was thought to be reflected in a diminishing body size, smaller brains and change in the horn forms (in animals that have horns) or a shortening of the jaw (in pigs and dogs). Scholars of domestication attributed the change in body size to early animal keepers purposefully selecting for smaller animals because they would be easier to handle. The shrinking of brain size, they thought, was due to animals under human care experiencing less brain stimulation and essentially having a more boring and less challenging life. They no longer had to run away from or defend themselves against predators, while humans also relieved them of the need to search for food. Both kinds of morphological changes, smaller body size and smaller brains, would have taken a substantial amount of time to become manifest.

There were no skulls among the bone remains which could provide information on brain size, but I performed a diligent metrical analysis of those goat long bone fragments that were measurable. The median size of the bones did indeed decrease over time, as was to be expected under a process of domestication. But surprisingly, this was entirely due to the one-sided elimination of large specimens, while the minimum size remained constant. If both male and female goats were experiencing a size decrease, then the minimum size should also have shifted, but this was clearly not the case. The only interpretation that made sense to me was that the much larger male goats had been eliminated from the picture – that is, they no longer reached full size because they were slaughtered before reaching adulthood. So, there was no actual size diminution of the entire goat population, just a change in the proportion of adult males versus females.[10] Was this sufficient evidence to declare the goats of 'Ain Ghazal as domesticated?

I was helped in making sense of the evidence by spending much time with an extended Bedouin family who belonged to the Amareen subgroup of the Huweitat tribe near Petra, the caravan city carved into red rocks in the south of Jordan.[11] They had pitched their tent at a scenic location amid huge boulders, and very close to another famous archaeological site from the New Stone Age called Beidha, which coincidentally also documents goat domestication.

Staying with a modern Bedouin family was a class in 'Living with goats 101'. Aisha, the eldest daughter of the house, had the chore of taking the fifty-odd adult goats out on their daily grazing rounds on the surrounding slopes where they browsed on wormwood, saltbush and aromatic herbs. While their mothers were out browsing, the goat kids had the run of both the men's and women's sections of the tent, jumping and dancing on its roof, knocking over cooking pots and pans, stealing food, playing catch, and generally being a nuisance, which provided ample entertainment of cartoon quality. They also served as playmates for the human kids of the household; they willingly allowed themselves to be carried around, cradled and cuddled, being much more responsive and haptic than dolls. While this was going on, the grandmother remained unperturbed and kept busy spinning the black goat hair into thick yarn with a drop spindle and then using a simple ground loom to weave this into new strips which could replace worn-out parts of the tent. Fatma, Aisha's mother, milked the goats in the evening and processed the milk into butter, using goat skins as a fermenting vat and churn. Sometimes she made balls of hard cheese, known as *kurut*, which were sun-dried on the roof of the tent. When guests came, Mohammed and Ibrahim would slaughter a young male goat, which would be chopped up in its entirety and cooked in yoghurt to make *mensaf*, the traditional feast of goat meat served on a bed of rice. The men also periodically sold male goats to cover household expenditures and buy tea, sugar and flour. Basically, the family goat herd was composed of females with offspring and only one adult male goat that had the privilege of ruling over this harem.

Such a composition makes complete sense, as there is no point in keeping male goats after they have reached full size. Once grown up, and even earlier, they engage in macho rivalry, try to eliminate each other, and are injured in the process. Nor are they going to put on any more weight, so there is no point keeping them, and the only sensible thing to do (from the human perspective) is to sell or eat them. This was exactly what was reflected in the bone samples at 'Ain Ghazal. In the early stages of occupation, goats had been hunted, which had resulted in both male and females entering the bone profile. In the latter stages, goats were husbanded. No actual size change had happened, only a shift in the population structure to the one I observed among my Bedouin friends' herd.

Experiencing the intimate relationship between the Bedouin family and their goat herd helped me envision how domestication may have happened. Very likely it was the women who drove the integration into the household: men brought back live goats from their hunting expeditions to keep as stock on hooves to be slaughtered when needed. Women looked after these goats, which would be tethered near their houses. Sometimes a goat gave birth, children played with the kids, and women nursed goat kids who lost their mothers, as is the practice among some non-Western cultures. They would have also accepted the help of nanny goats to feed their own infants.[12] Inevitably, goats were imprinted on humans and vice versa, creating a deep bond between the two species and resulting in what anthropologists now call 'hybrid households'. Although it was earlier presumed that 'domesticates' were initially used only for meat and that milking started thousands of years later in a 'secondary product revolution', it is now thought much more likely that utilisation for milk was right there at the beginning and a major mechanism for bonding.

I worked as an archaeozoologist in the 1980s. Since then, anthropologists have thoroughly revised their definition of domestication and moved away from the anthropocentric view that equates it with the control and mastership of animals by humans. Fieldwork with reindeer herders in Siberia, with multi-species pastoralists in Mongolia and among camelid keepers in South America has brought about this reconsideration.

Due to its long winters and dry climate, Mongolia is the one country whose economy was once almost exclusively based on pastoralism, and even today more than 80 per cent of the country's agricultural production is derived from pastoralism, which is practised on 72 per cent of its territory. The five most important livestock species are horses, camels, sheep, goats and cattle, although yaks and reindeer are also kept in northern pockets of the country. Without any doubt, herding is the foundation of its national identity.

Australian anthropologist Dr Natasha Fijn lived among Mongolian pastoralists in the Khangai Mountains and their herds of cattle, horses, goats and sheep. She deems them a 'hybrid multi-species community' and observed that the animals are rather autonomous, yet they still seek the companionship of their people or 'human animals'. She views humans as part of the herd in which they take the dominant position, but the relationship is based on trust and only functions if that trust is upheld.[13]

In Northern Asia in general, herders by no means seek to control their animals totally, and instead value the autonomy of their herds and their ability to fend for themselves. They encourage animals to graze on their own, without supervision, at least during part of the year. But they also strive for animals that cooperate and collaborate when needed, aiming for a fine balance between these somewhat opposing trajectories.[14] Camel herding in India's Thar Desert operates along similar lines: during the dry season, the camels roam around on their own to forage on dispersed vegetation that springs up at different locations, depending on where rain has fallen. They are collected in a communal effort during the rainy season and then efforts are made to build up a bond with them. This bond must be nurtured continuously. Anthropologist and fibre expert Dr Penny Dransart from Aberdeen, who worked for many years with Aymara llama herder-keepers, notes that the taming of camelids was not something that happened once several millennia ago, but that effort must be made to continuously renew the bond. 'The *wayñu* ceremony, an elaborate ritual during which the sexual maturity of young animals is celebrated, serves to reinforce the close relationship between animals and people.'[15]

If you think about it, the mixture of autonomy and closeness, and the constant work on the relationship, is a recipe for successful partnerships of any sort, be it between couples, organisations or business partners. It is also a strategic way of marshalling brains with complementary knowledge and skills to achieve a common goal. Conversely, if you subdue animals (or people, or even 'Nature'), you lose out on their abilities and have to expend a lot of energy upholding subservience.[16] And it also involves a lot more labour, feeding and servicing of animals, rather than delegating that responsibility to them.

In the nascent field of human-animal studies, the term 'domestication' has now been replaced by 'co-domestication'. It is no longer seen as a unidirectional process, but instead as a mutual arrangement. Researchers have begun to analyse the phenomenon from the perspective of the animals and muse that domesticates benefit from it just as much as the domesticator. They regard the whole notion of domestication as a Western concept. Some anthropologists have gone even further and started to scrutinise the perceived dichotomy between culture (the realm of human making) and nature (the world of animals, plants and geology) and natural forces. Life does not lend itself to such simple polarisation.

---

But what happened at 'Ain Ghazal – why did the place experience such a dramatic decline in fortune from a cultural hotspot and centre of refined living to an assemblage of shacks that was eventually abandoned? The excavators believe that the 'Ain Ghazalians ate themselves out of their initially bountiful environment, situated at an ecotone – a transitional zone between two biological communities, in this case between the steppe and the forested hills. Their predilection for plastered floors, plaster statues and plastered skulls was energy intensive. Making plaster requires high temperatures and lots of wood. So, they started cutting down the Mediterranean oak trees that hosted the wildlife they had hunted earlier. They increased their reliance on goats, probably not only for meat, but also for milk. Goats are wonderful animals that cannot do much damage to grown trees, but they do love to nibble on young and tender saplings, preventing the rejuvenation of forests. As populations of both people and livestock grew, tensions arose between crop growing and inquisitive goats infiltrating the fields. In order to protect the harvest and to access forage for the goats, some families took to seasonally wandering with the goats into the steppe/desert to the east where winter rains created lush vegetation. While 'Ain Ghazal declined, small and seasonally occupied sites cropped up in the vast deserts to the east. Their dominant feature are round enclosures or corrals that are interpreted as some of the earliest evidence for herding, although in combination with hunting and other subsistence strategies.[17]

In essence, this is how pastoralism was born – from the need to keep animals moving to provide them with year-round access to forage and prevent them over-exploiting local resources and interfering with crops in the field.

This ecological imperative eventually led to the divide between what British explorer and diplomat Gertrude Bell in 1907 dubbed 'the Desert and the Sown', the dichotomy between the nomadic pastoralist people and the settled agrarian communities that is characteristic of the Near East. A similar pattern was identified by Owen Lattimore between the 'Steppe and the Sown', between Mongolian nomads and Chinese farmers which led to the construction of the Chinese Wall to protect the latter.

The first steps towards pastoralism thus happened about 10,000 years ago, according to current knowledge. We will never know exactly the details of what happened in prehistory. But there is no doubt that this alliance with herd animals was a transformative and revolutionary development for humanity. It opened up the opportunity to inhabit ecosystems and landscapes where previously people could not exist. It was a way of 'civilising' major parts of the world where it is not possible to grow crops. 'In one sense, pastoralists herding sheep and goats "domesticated" the steppe and desert, converting plant life not directly useful for humans into meat and dairy products,' says Gary Rollefson, the excavator of 'Ain Ghazal.[18]

The alliance with animals also marked the birth of an antithesis between a settled way of life that depends on crop cultivation and an animal-oriented, mobile approach to food production. A way of 'living lightly' on the Earth that precluded accumulation of too many belongings because of the constant need for movement. This resulted in a colourful spectrum of nomadic cultures that are ethnically very diverse but also remarkably similar, with their daily rhythms shaped by the needs of their animal partners. All of them depend on the ability of animals to transform waste to worth. Something close to magic is involved.

# Communication

*The camels trust me and I trust them. When I am together with them, I am not afraid.*
MADHURAM RAIKA, camel herder[1]

Herding only works if there is trust between humans and animals. Establishing this trust requires patience and communication skills across species borders. I learned this the hard way several years ago when we were suddenly saddled with a bunch of camels that had been rescued from slaughter by an animal welfare group and needed a home. Naively, I had thought they could just be added to the handful of camels we already owned and that were part of a herd grazed by an experienced Raika, Adoji. With his decades of expertise, he would have easily managed this, but unfortunately, just at that crucial junction, health problems on top of old age forced him to retire. Since we could not find anybody to replace him, it was suddenly up to us to look after the newly composed herd – several breeding females with their offspring as well as the 'rescue camels'.

Until then, I had only paid short visits to our moving herd in the mornings before they went out to forage. During those visits, besides drinking camel milk tea, the special perk had been to be nuzzled – kissed – by friendly female camels. The camels of the Raika very decidedly disprove the familiar image of the species as ill-tempered, obnoxious, stubborn, stupid and spitting creatures. The females radiate calm and serene dignity in combination with a curious and inquisitive streak that often gets the better of them. If you stand quietly among them, they cannot withstand the temptation to check you out gently, carefully approaching you, then bending down their long neck, taking a whiff through the long slits that are their nostrils and then deploying their very movable upper lip to rub your cheeks and give you a cameline kiss. Baby camels are not quite as refined and may prefer to nibble your ear or rummage around in your hair. But beware: if you extend your hand towards them, as you would for a dog or horse, it will immediately break the spell and make the noble camel turn its head away. Instead, behave as if they were the Queen of England: wait until you are approached and don't touch. Camels will only interact with you on their own terms and that's part of what makes them so fascinating. Forget swimming with dolphins or espying a tiger on a jeep safari. For me, there is no bigger thrill than a tête-à-tête with a well-bred camel.

I have always been filled with wonder at how a couple of Raika herders could move around with, say 50 camels, and make them behave like an orderly school class waiting at a zebra crossing. They respond to verbal 'commands', or rather gentle encouragements, to disperse and browse, to assemble and move on, to catch up with the rest of the herd or to wait for their turn at a watering point. This is not unique to the herds of the Raika, as Bedouin camel herds are exactly the same.

But this pleasant and cooperative behaviour cannot be taken for granted. It's invested in the herder and is the result of skilled and purposeful relationship building, often over generations. This herd-building process had not happened with the camels we now had at hand. We did not have a functional herd, but an assortment of individual animals who were nervous and unhappy. It was an utter disaster. Because they did not know us, they did not trust us, and we did not have the communication skills to change their minds and convince them that our intentions were honourable. None of the camels was inclined to show any of the intimacy that I had revelled in earlier. Far from it, I could not even get close enough to handle any of them. If I came within a metre, they would turn around and walk to a safe distance, from where they would look down at me with a supercilious stare, saying something like: 'Who are you? I don't know or trust you!' Taking a cue from their mothers, the babies jolted away awkwardly whenever I appeared on the scene, as if I was a bogeyman. The worst were three young males who were always on high alert. Suffering from mange, a highly contagious skin disease, they urgently needed treatment, but capturing them for this purpose took hours every morning and involved some kind of lassoing that only enhanced their fears.

The rescued camels presented a different set of problems. They were docile, apparently having pulled carts in their previous lives. But they were neither interested in the company of other camels, nor did they know how to browse from a tree. They were used to having feed placed in front of them. They refused to go out grazing and integrate with the others. We had to keep them in a paddock on our campus and purchase expensive feed for them.

When we finally found a herdsman, this did not improve the situation much. For many years, he had worked as a village cow herd, but he did not have a knack for camels, although he craved their milk. Otherwise a sweet man, he was short-tempered and instead of talking to the camels in a friendly and soothing way, he frequently yelled at them, making things even worse. The status of our 'herd' was a huge embarrassment, and when a group of camel herders visited us from Jaisalmer, they just shook their heads in disbelief, remarking that 'We can see that these camels are not happy.'

Some months later, our herdsman was injured rather severely by a camel and went home to recover. The story went around that the camel had crushed him when he wanted to milk her. Eventually, we struck gold and found Madhuram from a long line of highly reputed camel and cattle breeders. Because of a lack of income, his family had already sold most of their camels and Madhuram had had a stint working in Mumbai which he hated. Happy to be able to work with camels again, he quickly managed to rebuild their trust and, within months, some of the young camels had become cuddlers, sneaking up to unsuspecting visitors and using their long, flexible necks to stick their spongy noses into their faces.

This episode drove home a crucial point: the Raika camel herds that I had always admired for their smooth manageability are the result of ties across a species boundary which need careful nurturing to be maintained, just as the taming of camelids in the Andes is a process that must be constantly renewed. It takes effort and skill to manage a herd, so that it functions as a single organism and does not splinter into different groups with animals wandering off. A herder must be a trusted leader whom her or his animals will follow. Herding systems only work if there is a positive chemistry between people and animals. As in any marriage or relationship, this rapport is not something that exists automatically, but that must be created and then continuously nurtured to be kept alive. It requires building confidence through consistent behaviour and positive feedback, regardless of species.

Here is what Saverio Krätli, who spent 18 months living among the WoDaaBe pastoralists in Niger, says about their relationship with their herds:

> The cattle bred by the WoDaaBe know nothing of enclosures, follow their herder of their own accord (rather than requiring to be herded from the rear) and it is common, in the bush, to see entire herds controlled by one or two young children waving only a twig. Indeed, although sophisticated and intensive, the WoDaaBe herd management is so smooth and light-handed that it appears, from the outside, as if the Bororo zebus bred by the WoDaaBe were actually committed to 'cooperating' with their herders. Behind such an impression there is, in fact, a characteristic 'attitude' of these animals, the development and maintenance of which is a key aspect of the WoDaaBe breeding/production system.[2]

So how does the communication between herders and animals actually work? Herders communicate with animals through sometimes barely discernible body movements and, importantly, by voice. They use certain sounds for encouraging their animals to disperse and forage, to assemble, to separate mothers and young, or to drink in an orderly fashion from a water source without crowding each other out. Sometimes they sing songs or use musical instruments like a flute, alphorn or violin or even a radio to communicate with them.

American ethnomusicologist Kip Hutchins spent some time with herders in Mongolia's Dundgovi during the short spring period when animals are weakened by the long winter and the weather is tempestuous. This is the birthing season and, due to frequent dust storms and night frosts, it often happens that newborn lambs are orphaned and require a foster mother. To persuade a ewe to accept a baby that is not her own, the herders sing to her while embracing her and the foster lamb, a practice called *toiglokh*. This procedure is repeated for a couple of days until ewe and lamb have bonded. The singing calms both mother and child, making them comfortable and instilling trust. *Toiglokh* and singing to other types of livestock are the foundation of traditional Mongolian music and vocal performances. They are thought to be instrumental in putting both human and non-human animals on the right path in life and creating social connections.[3]

Another example of using song to communicate with herd animals comes from Scandinavia where women traditionally spent three to four months every summer with their family's cattle herd in faraway mountain and forest areas to make use of pasture. In this system, which is known as shieling, the women used their voices as well as horns to communicate with cows over long distances, using different sounds for different messages. If the animals were far away, the women made arcs of sounds with a few strong notes that carried over distances of 3–4km (2–2½ miles). For goats they used other more playful sounds and another range to scare away bears and wolves. While this type of herding has now virtually disappeared, the same vocal techniques have become the basis of Scandinavian folk music and are learned by young concert singers to increase their range of vocal expressions.[4]

In Switzerland there were special songs for coaxing cattle to give milk, while the African Somali sing to their camels all the time, except while milking as the camels do not want to be disturbed then. They equip their wards with bells, so they know where they are when visibility is limited – in the fog, in the dark, or in the forest. The bells make different sounds and some pastoralists fine-tune them, so they resemble an orchestra playing a symphony.

Naturally, this is not a one-way conversation, but a dialogue. Animals also communicate their intuitions and emotions – fear, thirst, longing for their offspring, excitement, impatience, lust – through their voice and behaviour. They have a better sense of dangers, intuit the presence of predators and anticipate extreme weather events and earthquakes. They smell where rain has fallen and move in that direction to avail themselves of fresh vegetation. They are a source of intelligence that is invaluable to their keepers.

And a good herder follows his animals' advice.

The late anthropologist Paul Riesman lived among the Jelgobe subgroup of the Fulani pastoralists in what was then Upper Volta (now Burkina Faso), in West Africa, and noted that they could not live permanently around the mosque of their charismatic Islamic leader because the cattle refused to do so, which 'is a reason against which, for the Fulani, there is no argument. Cattle are considered to be one of the most intelligent species of animal; it would be crazy to do something the cattle do not like.'[5]

The Nenets are reindeer herders in the Yamal Peninsula, in Russia, who migrate up to 1,200km (750 miles) annually between the tundra and the taiga. Anthropologist Florian Stammler describes how they read the behaviour of their animals and act accordingly:

> Humans follow them throughout a 24-hour cycle of escaping from mosquitoes against the wind, feeding,

> drinking, and resting. Resting at the human camp site, it is the animals that decide when the resting time is over, and they slowly stand up and trot away to begin the next cycle. The herder structures the rhythms of his life according to the will of the animals.[6]

Stammler also describes how reindeer and people collaborate in choosing the best route to travel, with the reindeer deciding whether passages, for instance through boggy areas, are safe by testing the ground with their front legs. A sleigh driver will carefully watch his animals' behaviour, indicating to them the general direction in which he wants to go, but leaving it to them to figure out the detailed route. He also lets the reindeer resolve where to break the journey for rest and a quick graze.

When Madhuram first came to us to herd camels, he did not know the local territory. So, he had to learn from the herd. Instead of leading them, he let them be his guide and show him their favourite haunts: stands of especially tasty shrubs, places for taking a sand bath and natural watering points. He came to understand the area through their eyes. And after a couple of months, they insisted on moving away from the forest into the open fields, where some of their favourite vines and clinging plants that had proliferated during the rainy season had now matured. Instead of heading towards the forest, they turned the other way towards the open croplands. Gradually, as he learned, he was also able to guide them.

Often, herders make use of a third-party animal to manage their herd better and improve communication. In traditional circum-Mediterranean sheep culture, goats are used to manage the flocks more easily. They serve as leaders to keep the flock moving, as sheep tend to huddle together and refuse to move on if they cannot see a lead animal.

In Anglophone countries and many European countries, dogs are used for both herding and for protection against predators. In Romania, where bears and wolves pose a danger, future guard dogs grow up with the sheep and are fed with them, resulting not only in a deep bond, but also associating the flock with food. Shepherds think that the dog looks at the flock as prey that it has to protect from wild animals and other threats.[7]

In the Mediterranean area, and in the Carpathians in Romania and among some Afghan groups, herders make use of bellwethers as a kind of communicator between them and the flock. Bellwethers are castrated males that are imprinted on people and trained by them from an early age to follow their vocal commands. They then become the leader whom the whole flock follows. Sylvain Perdigon, an anthropologist from the American University in Beirut, studied the practice among Lebanese shepherds who migrate between winter camps in the Beqaa Valley and the mountains in summer. Male lambs selected to become a *meria*, or flock leader, are immediately removed from their mothers after birth and kept in a special section of the tent to maximise their time with the herders. For the first two or three months, they are fed with milk, then later they are given dinner scraps. At the age of four or five months they are put back into the herd, where they are at first unhappy, not knowing how to interact with the flock. They are then trained to walk behind the donkey by being bottle-fed under its belly. Normally there are several merias in the flock, they are not shorn, and they take it in turns to wear a bell. They have a calming and grooming effect on the flock and do their work for six or seven years, while the shepherd tries to uphold the close bond by regularly feeding the bellwether titbits. Not all ram lambs selected to become merias have the right disposition or develop the required close relationship with the shepherd.[8]

# Transformation, or the Difference Between Herding and Farming

*The really magical thing is that from grass we make meat. We add value to things nobody else can add value to. From garbage we make food. We protect air, water and soil, and the more there are of us, the cleaner the air, water and soil. Who else can say that about themselves?*

KNUT KUCZNIK, German sheep pastoralist[1]

I am very fortunate to live at the intersection between nature and agriculture, between 'the Forest and the Sown', on a tiny promontory at the foot of the Aravalli Hills. This chain of peaks, which meanders through Rajasthan from northeast to southwest, arose some 2,000 million years ago when the tectonic plates of India and Eurasia clashed and merged. In the east, the view from my house is dominated by the Kumbhalgarh Fort, which presides majestically above the Godwar area. The wooded hillsides below it host a burgeoning population of leopards and a few wolves; they have been declared a wildlife sanctuary and the option of re-introducing tigers is being discussed.[2] Towards the west, my house looks down onto a patchwork of small farms. Due to the runoff from the slopes, there is a narrow strip of land that is richly fertile and cultivated intensively. During some months of the year, I can always hear the rhythmic tuk-tuk-tuk of the diesel engines pumping out water to irrigate crops of mustard, sesame, corn and wheat. After a string of years with good rainfall, last year there was a bit of a drought, so the wells have run largely dry and a huge and noisy drilling truck is making the rounds to add a few metres of depth to them.

For the scrubby and drought-adapted native vegetation, the lower amount of rainfall is no issue. Every morning and every evening, at least three mixed herds of goats and sheep – each of about 50 heads – pass by my house. They come from the nearby village of Joba for their daily grazing foray in the forest and are all led by Raika wearing their signature red turbans, with water bottles swinging from one shoulder and balancing a long bamboo stick on the other. During the birthing season, they often come back home with a newborn lamb or kid draped around their neck. The herds quickly march through the fields in a tight formation, but as soon as they are past my house, they fan out into the prickly shrub forest, which is dominated by various acacia species so spiky that your clothes inevitably get torn if you venture off the track. The sheep, vacuum-cleaner-like, graze with their noses close to the soil, while the goats daintily search for pods and edible leaves higher off the ground. Of course, as it's a protected area, this grazing and browsing is against the rules, if not to say illegal, but that's not the (main) point here.

It's a peaceful pastoralist scene that will evoke different reactions depending on the onlooker's background. For an urban visitor it's a nice selfie opportunity with an exotic-looking native. The forester will say, oh no, these goats are destroying my trees, I will fine the owner. An animal scientist might think, this is so backward, are these goats getting a properly balanced feed? A photographer will revel in the colours of the turban, the moustache of its bearer and the speckled pattern of the goats.

But there is something much more profound going on here: the transformation of thorny plant fibre into pounds of nutritious protein food. The generation of a much sought-after and increasingly rare resource for farmers: organic manure. Sustenance of leopards, wolves, hyenas and other endangered predators, for whom sheep and goats are a preferred prey. Propagation of the acacia trees whose seeds must pass through a ruminant stomach to germinate. And, in stark contrast to the efforts of the farmers who require huge amounts of water, diesel, fertiliser and other inputs to reap a harvest from the soil, our three Raika, in an act of biomimicry, require only sunlight and mobility for carrying out their food-production activities.

---

There is no way around it: farming is – almost always – in conflict with nature. A farmer starts out by clearing the land and putting an end to its native vegetation, then prepares the soil by ploughing, harrowing and fertilising it. The next step is to put in seeds of her or his choice, replacing the original biodiversity with a monoculture (almost always – sometimes there are polycultures where more than one species is grown). Unless it's an organic or a low-input farmer, the crop is then protected with several applications of chemical fertilisers, weedicides, pesticides and fungicides. Finally, the crop is harvested and brought back to the barn or to the nearest silo where it is stored for onward transportation to a processing unit that will turn it into an edible product. In richer countries, these activities are performed by tractors and other machinery that require fossil fuel. In consequence, agriculture is the largest contributor to biodiversity loss and the second biggest contributor to climate change.

Now imagine a different model of food production. The land is not cleared and the native vegetation remains in place. The earth is not tilled. No applications of chemicals of any sort are required. When it's harvesting time, a solar-powered combine harvester is deployed to harvest parts of the perennial plants. It has an attached processing unit that converts the harvested biomass into two components: edible food and organic fertiliser. The food is dispensed, if required, but can also be stored until needed without any danger of it going off. And there are other perks. For one, the solar-powered harvester intelligently responds to voice commands. Secondly, it is self-replicating – you never have to buy a new one. Thirdly, it is modular, as it can be disassembled into smaller functional units, according to the size of the land that is to be harvested.

You may think this sounds great, but utopian. Yet it is what pastoralists have always done and what they have perfected: pooling their intelligence with that of herd animals to harvest and process surplus and biodiverse biomass – native vegetation or crop aftermath – and to do this in a natural way without a need for fossil fuels or chemicals of any sort. The herds are not just solar-powered harvesters; they are also processors, transforming biomass that plants have generated by means of photosynthesis into milk, meat and other products necessary for human life. They do this with the aid of billions of congenial bacteria that live in their stomachs and guts, and which break down human-inedible cellulose, the most common building block of plant material, into molecules that they then synthesise into muscle, skin, hair and milk.

Ecologically, pastoralist herds aggregate solar energy that plants have captured, converting it into protein in the form of meat, milk and fibre. Simultaneously, they fulfil the role of a refrigerator or freezer, they are a stock on hooves, a 'live-stock'. Flocks and herds respond to the voice of their herders better than a Google voice assistant because they are not pre-programmed. They reproduce on their own, so even if you sell some specimens, your herd normally replenishes itself (except in the case of a climatic emergency or disease outbreak). Finally, herds can be subdivided into smaller units depending on the resources that are available.

Ingenious systems like this operate all over India, from the Himalayas in the north to Kanyakumari at its southern tip, and from the Thar Desert in the west to the Bay of Bengal in the east. The country is virtually choc-a-bloc with herding cultures that rely on animals walking to their varied feed, which is composed of harvest waste and natural vegetation. Totally independent of fossil fuels, these human-animal partnerships make food 'out of nothing'.

Yet these systems, although they are in plain sight, remain unseen and unvalued. They usually do not register with livestock experts trained in the Western mould. And if these experts do take note of them, they despair about their backwardness and try to replace them with modern ways of production that include high-yielding breeds and as much high-tech as possible.[3] Intriguingly, despite being so 'low-tech' with regards to animal husbandry, India is a superpower in terms of livestock output. It ranks first in milk and sheep/goat production, besides competing with Brazil for the dubious epithet of being the world's largest 'beef' exporter. Indian 'beef' is actually mostly buffalo meat which is internationally subsumed under the 'beef' heading.

Fortunately, this stance of ignoring the actual status quo is now being challenged by an expanding group of activist researchers who recognise the value of the traditional, yet constantly evolving, systems and have begun to document them. In the process, they regularly keep discovering new ones that burst our confined horizons of what constitutes animal husbandry, such as duck pastoralism, donkey nomadism and pig herds kept for fertilising fields.

A jewel among this group of engaged scientists is Dr Balaram Sahu, a veterinarian from Odisha who is also an accomplished poet. I first met Balaram at a conference in Delhi where he presented a hauntingly beautiful video entitled the 'Night Queen of Chilika'. Chilika is the name of a 'lake' or rather a brackish lagoon on the east coast of India which is famous as a stopover for millions of migratory birds on their flights from Siberia to Africa. 'Night Queen' is an epithet for a special breed of scimitar-horned buffalo that swims in the lagoon at night to forage on its saline underwater vegetation. In the morning, the black behemoths return to their owner's house to deliver a few litres of precious milk and dung. Yet their relevance goes far beyond that mundane job: they are also an essential feature of the Chilika ecosystem, as their manure sustains its significant fish population. Alarmingly, this had become threatened by the government's efforts to replace the – in their eyes – unproductive native buffaloes with the popular Murrah breed which produces two or three times as much milk. Unfortunately, this higher-yielding type did not go for nightly swims, but stubbornly stayed on land. As a result, the fish population was deprived of the nutrients that had allowed it to thrive, which then affected the livelihoods of local fishing families.

When I sought out Dr Balaram during the tea break to congratulate him on the film and find out more, he immediately invited me to visit. Just before we were herded back into the auditorium by the conference organisers, he added the somewhat mysterious remark: 'Please also accept my invitation to be a visiting professor at my *Pathe Pathshala*, the roadside university for livestock keepers.'[4]

Intrigued, I booked a very early morning flight to Bhubaneshwar, the capital of Odisha, as soon as I had a few spare days. In the meantime, I read up about Odisha and its proud history. Earlier known as Kalinga, it was from here that Buddhism spread under its king Ashoka. Besides impressive Buddhist sites, it can boast of two of the most important Hindu edifices in India: the Jaganath temple at Puri and the Sun temple at Konark. Today, Odisha tends to be associated with poverty and tribal unrest. If measured by per capita income, it is indeed one of the country's 'poorest' states and one of the least developed in terms of access to education, healthcare and other services. This contrasts with the richness beneath its soil: aluminium, iron ore and bauxite, highly coveted by multi-national corporates for producing steel and manufacturing weapons. This juxtaposition of material poverty with extractive practices by some of the richest corporations in the world has caused tension and turmoil: some of the Adivasi tribes are putting up a spirited struggle for their ancestral lands.

An excited Balaram was awaiting me at the airport with a big bouquet of flowers, as if I was visiting royalty. His face crinkled into a wide smile, he ushered me to his small white Maruti car and proceeded to explain the programme he had planned. 'We have to hurry to get to the village before the buffalo go out grazing,' he said. While he was expanding on our programme, I caught a fleeting glance of a series of huge billboards by Vedanta, one of the mining corporations, trumpeting its efforts and benefits for women, children and rural development.

'We will first go to Chilika and then officially inaugurate the *Pathe Pathshala*. The director of Animal Husbandry is also coming.'

'But what is the *Pathe Pathshala*?' I asked.

'*Pathe Pathshala* means "university on the move" or "roadside university". It is my way of providing training to poor livestock keepers in how they can treat their animals with local herbs and without incurring expenditure. They don't need to – and can't afford to – buy expensive medicines when there are so many traditional treatments. I have documented them all in my book,' he said and handed me a volume with detailed drawings of medicinal plants and descriptions of their therapeutic value.

'You see,' he said, 'although there is a surplus of food in Odisha, and although the state sits on a reserve of grain, rural people still often go hungry. Livestock is a crucial source of food for them, but many of the new technologies that are promoted, such as antibiotics and artificial insemination, do not reach the common people, are too expensive or are not relevant. The people should be the target, not the technology as such, so in the *Pathe Pathshala,* we build on what people know and what has proven successful in keeping livestock healthy. It provides knowledge at Peoples' Place, in Peoples' Time and in Peoples' Language. And the people already know so much: about the medicinal properties of wild plants, about indigenous strains of grain, of biological ways of pest control, the nutritional values of pulses and vegetables, about traditional ways of preparing foods, about biodiversity in general.'

Engaged in lively conversation, we drove south from Bhubaneshwar on the four-lane National Highway. The Chilika lagoon is a pear-shaped water body whose extent varies from 900–1,200 square kilometres (350–460 square miles) between seasons and is separated from the Bay of Bengal by only a very thin finger of land which is submerged during the monsoon. Chilika is one of India's largest wetlands and the nation's most extensive waterfowl habitat, where 153 species of birds have been identified.

After about a hundred kilometres, we took a left turn onto a potholed, one-lane road that skirted the lake until we reached the first village. In the midst of it was assembled a herd of several hundred buffaloes. They were standing there quietly and unemotionally, as buffaloes do, and in their typical buffalo posture with their heads slightly down, their backs straight and their tails gently swishing. The place was like a beehive in which the uniformly dark brown buffaloes formed the honeycomb and people were the bees. Some were milking, but most of the action consisted of women industriously scooping up buffalo manure into metal bowls that they lithely carried away on their heads to what appeared to be individual processing sites. Here the dung was mixed with some straw, patted into flat discs and then plastered against the walls of surrounding houses or any other flattish surfaces for drying. The dung patties created intriguing patterns in which the hand imprints were like the signatures of the individual artists. I was absorbed in the symmetric artistry of it all, but then Balaram drew my attention to a young buffalo that was wearing a kind of apron made from hessian cloth hanging in front of his genitals.

'This is one of the Murrah buffaloes that the people don't want to breed,' he said. 'I think it has been given by the government for upgrading the Chilika buffaloes.'

'Hmm. Interesting,' I said.

'Can we ask some people what they think about the introduced Murrah bulls?'

Balaram scanned the buffalo dung patty assembly line for an informant who was not too busy to answer questions. He zoomed in on a frail elderly man who looked friendly and wise and was holding the tail of a buffalo, while a child – maybe his grandchild – crouched under her and stripped milk into a metal bucket. A short dialogue followed which I obviously could not understand, although I noted a wry smile appear briefly on the man's face.

'He says that the government buffaloes are like a Maharani. They don't go out foraging like their own buffaloes. Instead, they wait for food to be served to them. For them the original buffalo is much more useful as they don't have to worry about feeding them.'

I would have loved to ask more questions, but suddenly there was commotion: the mass of buffaloes had started to move, and sluggishly but unstoppable, like molten lava flowing from a volcano, they poured out of their assembly place in the direction of the road on which we had arrived.

'Quick, quick,' urged Dr Balaram. 'Come with me so we can take photos and footage of how the buffaloes move.'

We ran ahead and posted ourselves at a strategic point on the roadside from where we could get profile shots of the buffalo cavalcade. I pointed my camera at the buffaloes crossing the road, concentrating on clicking at the right moment. Subconsciously, I registered a minuscule old lady in a very faded, grey-green sari hunched by the roadside a few metres away from us. Suddenly, out of the corner of my eye, I caught a rapid movement. The woman lurched forward in a predatory arc that would have done any feline proud to rapidly scoop up a steaming pile of manure which one of the creatures had dropped on the asphalt. Remarkably nimbly, she now walked off, clutching the booty to her chest, with what seemed like a triumphant spring in her step.

Puzzled why this old lady had to hunt for manure like a vulture for a dead animal, when the whole village was literally plastered with it, this was my first question to Dr Balaram when the buffalo procession had crossed the road and walked down a path towards the lake.

He had a ready explanation:

> 'Manure belongs to either the owner of a buffalo or the owner of the land where the manure drops. If the manure falls on the road, which is public property, then it belongs to nobody. So, this lady availed herself of the opportunity. Probably she does this every morning. She will dry the manure and then either use it as fuel for making food or maybe even sell it.'

We spent a few more minutes observing the buffaloes as they marched in the direction of the lake with their heads held down, tufts of rusty coloured hairs standing out from their sleek black coats like needles on a pin cushion, each one surrounded by an orange halo of morning light. Like a ground force on autopilot, they slowly stepped over a railway crossing before determinedly plodding on to the shore of the lake. Finally, they stepped into the water in an almost bored manner, gingerly setting one foot before the other as if afraid of getting wet, but eventually plunging in and floating off towards the horizon.

We ourselves had to rush to be in time for the inauguration of the *Pathe Pathshala* and there was no time to sit down with the buffalo keepers and understand in detail what we had just witnessed. The value of the manure that the buffalo generated after swimming and feeding in the lake during the night was obviously enormous if it spawned this almost industrial-scale activity. Of course, the dried dung patties were a valuable source of energy for cooking and for which no cash had to change hands. But were these pieces of manure art all used in the village or were they also being sold, maybe for ceremonial purposes in temples? Was the manure used by farmers as fertiliser? There were many questions in my mind that remained unanswered about this alchemy of converting the underwater vegetation of the Chilika Lake into fuel – in essence making fire out of water!

Dr Balaram and I have remained in regular touch over the years and some time later he called to tell me that he had discovered swine nomads – a small sensation since nobody had ever reported the existence of such people in India before.[5]

So, another trip to Odisha was in order. This time, we pointed Dr Balaram's car north and drove along the coastal plain and past some Buddhist stupas (mound-like buildings used for meditation; the name is derived from the Sanskrit word for "heaps"), until we reached the delta of the Mahanadi River, a fertile landscape excellently suited to rice cultivation. Near the village of Patamundai in the Kendraparha district, the pig nomads were waiting for us in the middle of a harvested rice field. I could make out about fifty small, brownish pigs of different sizes rummaging around for any rice corns that had escaped the harvest. It seemed like a systematic exercise of scanning a given area for something very specific: my first impression was that of a troupe of sniffer dogs hot on the trail of drugs. Or maybe closer to home, the pigs that search for truffles. The group was an extended family that greeted us in a friendly, although somewhat shy, fashion. We sat down in the shade of a pile of rice straw to talk; several women in colourful but faded saris and some of them with small babies in arms formed a half-circle behind the leader, a weathered middle-aged man wearing shorts and an orange scarf over his shoulder. After some mutual introductions, he led the discussion. Although I could not understand his language, it was evident from his manner that he spoke in a concise way and very pointedly. He emphasised that their pigs were entirely healthy and this had been confirmed by the veterinary department. During the recent (human) swine flu outbreak, some people had considered the pigs dangerous and questioned the need to keep them. Then he described how the pigs were eaten by everybody and how he had a continuous flow of customers who came to buy pigs for home consumption. They paid between 1,000 and 1,500 rupees for piglets and around 4,500 rupees for a grown-up one. Each sow has two litters per year, averaging about 10–12 live offspring annually, of which about half survive. This amounted to an annual income of around 20,000–25,000 rupees per sow.

From what I could see, it was an entirely positive system, or in development lingo, a win-win situation – the pigs were happy, searching out their food and utilising natural resources that would otherwise go to waste. The rice farmers were benefitting from the manure from the pigs. The pig breeders had a fairly lucrative income, and consumers had access to tasty and nutritious protein and an entirely natural product.

But what should be recognised as a model of agro-ecological food production that benefitted everybody had until then never been noticed and remained invisible to officials and animal scientists who are primed to value only white European hybrid pigs – animals that are kept in 'proper' housing and stall-fed with expensive and specially cultivated nutrients.

And so it goes that all over India nomadic livestock keepers walk with their herds to forage. In the north, the Van Gujjars are one of several groups that move with their herds up and down the Himalayas. They keep buffaloes with whom they have intimate relationships and who decide when it is time to move. But the forest department has now stopped them migrating, allotting them fixed housing where it is no longer possible to keep their buffaloes, which need to move with the seasons. On the west coast of India, there are Rebari and Jatt pastoralists who raise the Kharai camel – a breed that seasonally swims into the ocean to forage on mangroves, the growth of which is stimulated by being browsed and which protect the coast against erosion and flooding. In India's interior, there are Kuruba and Dhangar shepherds that feed the cities while at the same time helping to conserve wolves whom they worship.

Although, or maybe because, they manage that utterly desirable feat of producing food AND replenishing the environment, they are not under the purview or of interest to the government departments concerned with these issues. Foresters and environmentalists regard them as destructive of nature. Agricultural and animal husbandry experts deem them backward practitioners of animal husbandry that need to be taught how to raise livestock according to scientific standards. Public health officials and the general public worry that they harbour and spread diseases, although they personify the much-talked-about One Health approach that regards the health of people, animals and ecosystems as interconnected and which is strongly promoted by the World Health Organisation.

---

In December 2016, hundreds of pastoralists from all over India assembled in Delhi at an exhibition that celebrated their way of 'living lightly' on the Earth, together with their art, songs, food and products.[6] There were Bakkarwal and Gujjars from the Himalayas, Raika from Rajasthan, Rebari and Jatt from Gujarat, Dhangar shepherds from Maharashtra, Kuruba from Karnataka, Golla from Odisha, Kangayam cattle breeders from Tamil Nadu. The colourful convention vociferously rejected the suggestion that they should be classified as farmers. Instead, they insisted on having a separate identity as animal herders because they do not own land, do not till the soil, and don't grow plant crops. Instead, they team up with animals to make use of what is already there and upcycle it into valuable products, complying with agro-ecological principles.

The etymology of the word 'agriculture' gives us a clue. It is derived from the Latin verb *colere* – meaning 'taking care of, looking after' – and *agri* is the genitive of *ager*, which is Latin for 'field'. It literally means cultivating a field. But pastoralists do not cultivate fields. They husband herds or cultivate 'Fields on the Hoof', as the American ethnologist and missionary Robert B. Ekvall, who spent decades with Tibetan nomads, termed it.[7]

The range of eco-zones that they can access and the variety of vegetation or biomass they can metabolise is mind-boggling. In the Arctic, reindeer dig out lichen under deep snow cover, forage in birch forests, and graze the mosses and grasses of the tundra. Yaks feast on coarse grasses and small sedges that grow along streams fed by glaciers. Buffaloes feed not only on the underwater plants of the Chilika Lake, but also trudge out at night to nourish themselves on the Banni grasslands in Kutch. Not far from there, on the coast of Gujarat, camels seasonally swim out to small islands to feast on mangroves, whereas in the Thar Desert, they browse on spiky and thorny tree crowns high off the ground that they reach with their giraffe-like necks. Pigs kept in nomadic systems feed and fertilise harvested rice fields not only in Odisha, but also in Bangladesh and Southern China, while in the *dehesas* of southern Spain and in Portugal they feed on acorns dropped from four different types of oak. Sheep have an amazing ecological range, preferring herbaceous plants such as saxifrage, pea, stone bramble, heather and willow in Iceland, while in Southwestern Nigeria, they feed on 160 different species of lianas, herbs, grasses, trees and shrubs, belonging to 130 genera and 45 families.[8] Goats in Imo State of Nigeria have an even more diverse smorgasbord composed of 177 taxa of trees, shrubs, vines, herbs and forbs.[9] Cattle in the Sahel predominantly eat a diversity of grasses, with herbs and some tree foliage also thrown in. The South American camelids, alpacas and llamas forage on a huge variety of grasses, sedges, reeds and forbs. Keepers of Mithun, a semi-wild bovine, in the Yunnan Province in China, cited 142 wild forage, belonging to 58 families and 117 genera and including 61 species of tree / shrub fodder plants and 81 species of herb forage plants.[10] The North African Tuareg know at least 38 grasses and herbs as well as 24 shrubs and trees. The camel-keeping Rendille in Northern Kenya identify 47 camel fodder plants that they separate into bulk feed and those consumed in small quantities as a kind of herbal health supplement.[11]

Whether vegetation grows in well-kept inner-city parks and lawns or at the ends of the Earth, we can harvest most of it by means of herded livestock. 'Grass-fed' signifies natural livestock keeping and marks out products as healthy to consumers; it is in vogue all over the 'North', but herding is about much more than utilising grass: it converts literally tens of thousands of naturally occurring forage plant species into food. It thereby circumvents that stage of creating artificial uniformity – and eliminating biodiversity – on which the bulk of current food production depends. Currently, less than 200 plants are cultivated on any scale, with only nine (sugar cane, maize, rice, wheat, potatoes, soya beans, oil-palm fruit, sugar beet and cassava) making up 66 per cent of total crop production in 2014.[12] Through livestock we are able to infinitely increase the number of plants we can harvest and the range of micro-nutrients in our diets.

While normal agriculture is based on uniformity, stability and control, herding capitalises on variability.[13] Variability is a feature of plants themselves – their abundance varies from year to year, depending on rainfall. Their nutritional content also changes during their growth cycle and for livestock health and well-being it is essential, or best, that they feed on plants when their nutritional content peaks. While variability is the enemy of crop cultivation, commodity trade and industrial value chains, it is the friend of pastoralists, as it means there is always something somewhere to forage on.

This foraging knowledge – of both animals and people – is extremely valuable as it means knowing how to make use of natural environments without fundamentally altering their ecological functionality. For an ecosystem, there is almost no difference between whether it is grazed by wild herbivores or by domesticated herd animals that move through it. It is not forcefully raped of its natural vegetation through ploughing, the cutting of trees or chemical poisons. Its underground soil microbiota is left undisturbed and its insect life stays intact, which means that birds and reptiles will keep their homes. Predators still have plenty to eat, maybe more than before.

Herding means to ride piggy-back on the natural processes with which our planet functions and maintains its equilibrium.

It's as natural a food procurement strategy as hunting, with one important difference: there is no flight distance between people and animals – the latter do not run away when approached. It is a relationship of trust. Humans and livestock communicate with each other freely and frequently, having learned to understand each other's language.

# Movement

*To the ordinary observer passing through the country, it would appear from its desert-like aspect that the rearing of cattle would be an impossibility, but if enquiry be made and the cattle in and around the villages be examined, it will be noticed that they are in good condition although their food consists of nothing but what they pick up on the apparently dried up grazing ground.*

F.S.H. Baldrey[1]

It's early October, and Madhuram's second year of herding our camels. They are a contented lot now, in good hump, without visible traces of mange and with relaxed, meditative expressions on the faces of the females, while the latest crop of offspring, now six to nine months old, are friendly and curious with a dose of mischief. They have developed an internal hierarchy that allows them to get along without friction most of the time. They unanimously accept Madhuram as their lead animal and wait for his guidance and instructions, although this does not prevent them from playing the occasional trick on him. He and the herd are a team, resembling a school class with a respected, even beloved, teacher. Or, if we want to use a biological simile, they represent a functional organism that harvests a range of trees and shrubs and metamorphoses them into milk…and new life, as a number of the ladies are expecting to give birth in the next couple of months.

During much of the year, the Raika aggregate family herds into small groups to share work and for company. But now, at the tail end of the rainy season, Madhuram and his herder colleagues have split them into their individual holdings. They have positioned their herds in a string of locations at the edge of the forest. The camels stay in the field of a friendly farmer during the night and forage in the forest during the day. At this time of year, the wood is still lush with excellent browse, such as the white-barked acacia tree, which is in bloom now, a mid-sized tree with a crooked and gnarled trunk and feathery, green-greyish leaves, and crowned by clusters of cream-coloured florets which camels love. Madhuram's routine is to milk the lactating camels before dawn and, once the milk has been collected by Hariram, a herder friend with a motorcycle, he leads his protégés deep into the forest to grazing spots that will keep them occupied for a while. He then strides back to his temporary home and has a simple brunch of fresh flatbread and leftover vegetables from the night before. After puttering around for a bit, he takes off again some time later in the morning to go back to the place where he left the camels. From there he tracks them down by their footprints, watches over their browsing and then returns with them after sunset. This system allows the camels maximum time to feed and gives Madhuram a bit of free time to take care of domestic chores, such as washing his clothes, mixing some ointment to treat camel wounds or repairing the scrub fence.

Today Madhuram has breakfast as usual, chats to the farmer, crushes rock salt for feeding the camels the next day and then takes off to join his herd. When he reaches the spot where he left them, they are gone. So, like a pastoralist avatar of Sherlock Holmes, he scans the ground for heart-shaped camel footprints and dung. He doesn't take long to find these, but surprisingly, today they do not lead deeper into the forest; instead, they lead in the opposite direction, towards the farmland. He follows the tracks until he reaches a tarmac road, at which point there is no indication as to which direction they went. Madhuram takes out his phone and calls some of his acquaintances to ask if they have seen anything. Success! The camels have been spotted near a petrol station some two miles away and were last seen slowly drifting towards Sadri. Madhuram hitches a ride on the backseat of a motorcycle and, sure enough, just before Sadri, he finds the herd ambling along the narrow, thinly trafficked road and browsing on the neem trees that line it. Adding some emphasis to his admonishments by waving his long, bamboo herding stick, he tells the camels to turn around and move back towards the forest. Seemingly relieved to see him, possibly even feeling slightly guilty (although this may be anthropomorphising them), they obediently march the way they came until they are once again safely ensconced in the forest, away from speeding vehicles and other possible dangers.

When Madhuram laughingly tells me the story in the evening, he adds: 'It is the time for them to move towards the farmland. They know it is now the Hindu month of *Kartik*, when the cold season starts, and they no longer want to browse in the forest. They remember the road we took last year. Camels always remember every road they took.'

'But why,' I ask him, 'why did they decide to walk away, since there is still plenty to nosh on in the forest?'

'Yes, there is,' he says. 'But at this time of year there are a lot of vines and creepers that have grown over fences and hedges along the small back roads, and they really like to eat those. So that pulls them away.'

---

Animals are made to move. Unlike plants they do not have the ability to photosynthesise and cannot generate energy from sun and air, but instead must obtain it from plants or prey. Wild animals, including insects, fish, reptiles, birds and mammals, migrate seasonally in order to find better food or avail themselves of more favourable living conditions. In the case of herbivores, this can involve huge distances. For instance, caribou cover around 1,000km (600 miles) between their summer and winter grazing grounds in Canada, while in Eastern Africa, millions of wildebeest, zebras and other assorted hooved animals undertake an annual circular migration between the Serengeti in Tanzania and the Masai Mara in Kenya, following the rains and the growth of grass. Scientists call this phenomenon of shadowing the vegetation as it greens up sequentially over large areas, depending on altitude and rainfall, 'surfing the green wave'.

These wanderings are not genetically embedded in their DNA, but are the result of learning over decades or even centuries. They are a skill that animals transmit culturally from one generation to the next, with offspring learning from their mothers. As Madhuram notes about camels, migratory animals have an excellent spatial memory, a kind of Google Map of their terrain embedded in their brains. These mental maps also have a temporal dimension: they include information about 'opening times' – for example, at what time of the year greenery or other forage plants become available – and, according to Professor Brett Jesmer who led a team of researchers studying the phenomenon among bighorn sheep in the Rocky Mountains of the US, migratory animals time their movements accordingly.[2]

The once abundant but now very rare bighorn sheep stay high up in the mountains during the summer and move down to the foothills in the winter. Professor Jesmer and his team studied what happens when you translocate them to new areas and found out that the bighorn sheep stay put, at least initially. It takes time for them to understand their new environment and develop a mental map of the area. Only after a few years do some of them begin making seasonal movements, gradually increasing the amplitude to 40km (25 miles), or even more. Developing a full migration cycle can take decades for translocated animals and several generations of learning.[3]

Interestingly, many wild animal populations are only partly migratory, meaning some individuals migrate, while others don't. If you compare these two groups, the migratory animals are fitter, bigger and more fertile (having more twins). So, there is a net fitness benefit to moving with the seasons and 'surfing the green wave' because it means that they consume the plants at the time when their nutritional contents are highest.[4]

It is no different with the keeping of domestic animals, which mirrors the situation in the wild. Livestock that is kept on the move is healthier and provides higher-quality products. The Mesta, the Spanish association of shepherds founded in the thirteenth century by King Alfonso X of Castile (often referred to as King Alfonso the Wise), recognised this early on: in order to pamper and provide optimal nutrition to the Merino sheep that produced precious wool, they moved them from summer pastures in Castile and Leon to winter grazing in the Extremadura and Andalucia, gradually developing a huge network of drove roads, the *Cañadas*, which was 125,000km (78,000 miles) long and protected by royal decree.[5] (The more mundane Churro sheep were not given the privilege of seasonally migrating to greener pastures and only kept locally.) The practice of emulating 'green wave surfing' with domestic herd animals is known as transhumance.

High-altitude pastures, whether in the Alps, the Andes, the Himalayas or the Carpathians, provide extremely nutritious, herb-rich forage during the summer that boosts animal health and enables livestock to make it through the winter on hay and much less nutritious feed. From medieval times until the early twentieth century, transhumance was common in Iceland, Sweden, Norway, Ireland and the Scottish Highlands and Islands. Herds of cattle, sheep or goats were kept in the mountains or moors during the summer around seasonal settlements that were known as 'sel' in Iceland, 'bouley' in Ireland, and 'shieling' in the U.K. The purpose was to reduce grazing pressure around the main settlements and to let animals benefit from seasonally lush forage. In the shielings, animals were often tended by women and children who had the task of processing milk into butter and cheese. While this involved hard work, it was also a time relished for the freedom it provided. Shieling is now no longer practised, although in the Scottish Isles there are still old people who remember it, as I was told by Sam Harrison of the Shieling Project, an initiative to revive this ancient land management strategy. Located near Beauly in the Scottish Highlands, the project serves as a learning centre for outdoor living where children and young people can learn how to look after livestock, milk cows, weave baskets, build houses, make food from the produce they grow, and other skills. The intention is not only to provide a window into the past, but also to demonstrate sustainable ways of living for the future.

In the Burren, in County Clare in Ireland, about a thousand farming families send their cattle into the rocky karst landscape (a limestone region characterised by underground streams and sink-holes) during winter in a reversal of the cycle. Here the animals feed on its herb-rich grasslands and rest on the limestone platforms which have banked heat during the summer and now provide a place to lie down comfortably when it is soggy elsewhere. The practice, known as the Burren Winterage, produces not only pricy gourmet beef, but also harbours an array of rare plant biodiversity.

---

'When we take sheep on migration, they are more productive and more profitable,' say the Raika who keep sheep in the Godwar area of Rajasthan where I have been based for the last 30 years. Only small flocks can be sustained all year round in the immediate environs of a village. If you have a larger herd, then seasonal migration for 'greener pastures' is a must. But in the early twenty-first century, sheep migration in India is a huge challenge, as open spaces have disappeared, highways have superseded ancient passage routes through the landscape, forests have been converted into fortresses of wildlife conservation and most fields have been fenced off.

Still, some Raika sheep nomads persist: those people who cannot fathom living in urban areas, degraded to menial labourers when previously they were their own bosses. And who love working with animals. Even when enmeshed in an ever-tightening web of land used for other purposes, these independently minded spirits manage to navigate their passage and negotiate sequential access to patches of 'green', which in practice are mostly harvested agricultural fields with crop residues. To manage this and to protect from the dangers en route, the Raika form herding groups of 6–20 families that stay together for a whole migratory cycle. This multi-species conglomerate is called a *dera* and, besides the human animals, consists of thousands of sheep, a lesser number of goats (as wet nurses), dozens of camels, and sometimes donkeys to carry belongings, as well as a few dogs. It is guided by an elected leader, the *patel*, who usually has an assistant, the *up-patel*.

The basic herding cycle of the Raika of the Godwar area looks like this: for nine months of the year, starting around the Hindu festival of Diwali in late October or early November, they move southwards towards the fertile agricultural lands of Mewar and Madhya Pradesh to graze flocks on fields that have been harvested. While doing so, they fertilise the soil with organic manure. Farmers recompense this service with food and even money, well aware of the difference this makes to the productivity of their land and the size of their harvest.

During the monsoon, when crops are sown and grown in the fields, the Raika herds all converge on their home villages in Godwar. Here they spend the three months of the rainy season, from around 1 July to Diwali, which marks the beginning of the cold season. They cannot return to the farmland until after the harvest. This basic pattern varies from year to year depending on rainfall; in a drought year when the landscape is desiccated and does not burst into the usual steamy-sultry greenness, the herders may stay only very briefly and move out to the fields that the farmers can't cultivate under these conditions. In years with good rainfall, they might extend their home stays because there is plenty of biomass to graze on.

Migration requires leadership and a strong and internally cohesive group, not just of people but also of animals. It is physically tough, and even more so socially, for all the human and non-human members of the *dera* must stick it out together, through thick and thin, for the average nine months of a herding cycle. During this period, they never have a roof over their heads or a semblance of privacy; they have to brave cold and extreme heat, occasional downpours and storms, uncooperative administrators, irate forest officials, nightly marauders and speeding trucks that kill swathes of sheep. With the next catastrophe always around the corner, there are plenty of reasons for tensions to run high. In the face of this, the internal harmony of the herding collective must be maintained or the whole operation has to be aborted because it's all based on teamwork, and nobody can default on their duty. Being in a *dera* is an extreme situation, similar to that experienced by a submarine crew or on an expedition to a high mountain. Maybe even like being in a rocket to outer space.

But there are rewards for this risk taking, too. The combination of larger flocks with higher production means relative wealth. At the Sadri bus station, one sometimes sees small bands of Raika emerging from a rattling and tattered bus after an uncomfortable overnight trip. Their footwear shows that they walk for a living: heavy, beaky shoes made from thick leather that is enforced with metal studs and with soles as fat as a sandwich. What is most striking about them, and incongruous with their weather-worn appearance, is their gleaming jewellery: the men sport heavy golden pendants and massive golden adornments draped over their earlobes. The women wear golden and filigree, disc-shaped nose ornaments that are pulled to the side by a little chain tied to one of their ears. The spectacular jewellery identifies these travellers as shepherds who are wealthy due to owning large flocks. They have taken a brief leave from migration to attend a social function at home before heading back to their herds, which may be 1,000km (600 miles) away.

During the monsoon, the shepherds are settled in the relative comfort of their homes and, in good years, there is an abundance of greenery and biomass. Yet, it may still be the most difficult time of the year. As crops are growing in the fields, the only area where forage is available is the forested slopes of the Aravalli Hills. Historically, the herders held customary grazing rights which were bestowed on them by the erstwhile Maharajahs who ruled the area and were well aware of the economic importance of pastoralism. But since Independence, these rights have become progressively curbed, starting in 1971 when 610.528 square kilometres (235.73 square miles) along 85km (53 miles) of the western face of the Aravalli Hills was designated as the Kumbhalgarh Wildlife Sanctuary. Then, in 2004, these long-standing rights were abolished altogether by the Supreme Court of India at the behest of wildlife conservationists. The local NGO Lokhit Pashu-Palak Sansthan (LPPS) – Herders Welfare Society – that Hanwant Singh set up in 1996 to support the Raika and other herders, had put up a valiant struggle and initiated several court cases against this ruling, but without success. This turn of events led many of the Raika to abandon shepherding, with almost all young people having to seek work in the cities as lowly paid labourers, often working 16-hour shifts and earning a pittance compared with what they could make from shepherding.

---

Dark and ominous-looking clouds were racing across the sky. There had been a thunderstorm and an outpouring of heavy rain in the night, the atmosphere felt damp even in spirit, and the green of the trees and crops in the field seemed psychedelic in its intensity. Our jeep was splashing through puddles and bumping through potholes on the way to the hamlet where a *patel*, one of the leaders of the long-distance shepherds, lived. He had phoned in several times urging us to visit, so I tagged along with Hanwant Singh, the director of Lokhit Pashu-Palak Sansthan, and Dailibai Raika, who is one of the board members of the organisation, hoping to better understand sheep migration practices.

These two friends of mine are formidable characters with a record of passionately and fearlessly speaking up for the rights of herders, locally, nationally and internationally. Dailibai is a traditional Raika woman, covering her head with a red veil and wearing a wide swinging skirt and chunky silver jewellery as well as a sizeable golden nose ring. But she is also an activist who has travelled the world and spoken about the Raika situation at the UN and other international meetings in places such as Nairobi, Montreal, Berlin and Madrid.

She is an incredible orator in Marwari, her native language, and can put the fear of God into recalcitrant bureaucrats or obnoxious forest officials who want to stop the Raika grazing their animals.

Hanwant Singh is not a Raika, but belongs to the Rajput land-owning class, yet his heart is with the landless nomads. Having worked with the Raika and me since the early 1990s, he has come to be trusted by them to the extent that they call him whenever they have a problem and expect him to sort it out for them. He has helped them to take legal action and has developed a successful carrot-and-stick approach for dealing with decision-makers – polite flattery and courtship in combination with blunt requests for the right action and no qualms about uncovering corrupt officials.

Soon we veered off the bumpy tarmacked strip onto a gravel road on which our jeep slid and slithered along until we reached the Raika *dhani*, or hamlet, which sat against a hillside, tucked into the fringe of the forest. Protected by a canopy of old neem trees, it consisted of about six low-slung, brick houses, their roofs overgrown with webs of creepers and the foliage of flowering vines. Each building was surrounded by several sheep pens and shacks, with stacks of manure piled up next to them.

Dharmabhai, the Raika leader whom we had come to meet, was busy in the sheep pen. He was drenching lambs with mustard oil, assisted by his young daughter who was holding them with a firm grip, while he poured the yellowish liquid into their mouths from a spouted claypot. They quickly finished their chore, then set up a couple of string beds for Hanwant and me to sit on while Dailibai disappeared into the house to meet the woman of the household. There is strict gender segregation in Raika society, although this loosens up during migration. Fortunately, as a non-Raika, I am free to sit with both men and women.

The monsoon rains are always desperately awaited, and resuscitate the region and the entire Indian subcontinent, but they come at a price, for both people and animals. The constant wetness and pools of standing water favour the outbreaks of diseases. Parasites have a field day and buzzing clouds of flies and smaller insects follow you around. The sheep look bedraggled and their white coats turn into yellowish mats, because they sleep in their own manure. The smallest scratches incurred while grazing in the scrub forest fester endlessly, often growing into balloon-sized abscesses. And ironically, when everything is lush, green and dripping, the sheep have the least to eat because there is no space for them to graze legally.

Once we had drunk tea and exchanged pleasantries, we asked Dharmabhai how the sheep coped and whether they liked being back in their village. Dharmabhai explained:

'The sheep hate this wetness. They are much happier on migration. As soon as the rains stop, they really want to leave, it becomes more and more difficult to herd the flock back here in the evening, and they try to move south. There is a lot of tension as we cannot take off until the fields have been harvested, otherwise the farmers get nervous about crops being destroyed. We have to avoid conflict.'

'So how do you decide when to make a move?'

'Well, when we hear from our contacts that some fields are open, our *dera* gets ready to move. But it has become so difficult. Especially theft has increased. Then there is the traffic – a highway has been built on the road we have been migrating on. The trucks go fast and last year we lost about 20 sheep when somebody just drives into them. When we complain, nobody helps us. The police and administrators don't listen to us, or they want money from us. Some farmers no longer want us,' he sighed.

'So how do you form a *dera*?' I asked. 'Is it always the same families that migrate together? Are you all related to each other?'

'Well, we try to include only compatible people with a good reputation and to avoid troublemakers. If somebody has a history of troublemaking, then no *patel* will accept him in his *dera*. So, yes, it is often the same families who move together, but there is also some change from year to year. Sometimes people get too old and don't have a son to continue with herding or some wives don't like to go on migration.'

After a thoughtful pause, he continued with emphasis: 'But the biggest problem are the thieves. They come on motorbikes in the night and take away our lambs.'

Then he looked at Hanwant and said, 'That's why we want gun licences,' adding, 'We don't want to use guns, we just want to carry them and scare away the thieves.'

I gasped because I had not expected this turn of conversation, but Hanwant nodded his head. 'Very understandable. It is difficult, and I know that this is a long-standing demand of the community. But I will talk to our MLA (Member of the Legislative Assembly) and ask him to raise the issue in the Legislative Assembly. We will try and let you know.'

Dharmabhai appeared satisfied by the answer to his appeal, which seemed to have been the reason why he had wanted to meet. The conversation drifted to other topics. When we finally said goodbye to Dharmabhai he insisted that we visit the *dera* when it was on the move. As of yet, he could not say when they would leave, but he would let us know.

A few weeks later, Dharmabhai sent a message that his *dera* was to depart the next day, but I was not able to witness the departure. Nevertheless, I had frequently enough caught glimpses of the operations involved to have vivid images of it. How the sheep bleated, stomped their feet and refused to be herded back to their enclosures in the evenings. How Dharmabhai got confirmation from the priest for the exact date, how there were last-minute discussions within the families who would migrate and who would stay behind. How the group assembled and loaded the camels with mattresses and blankets on which were placed woven carry bags topped with charpoys onto which the small children are strapped. It is always women and girls who walk with the camels at the front of the procession followed by the individual sheep flocks. Eager to be on the move, the sheep joyfully show their happiness by rambunctiously jumping into the air. Some newborn lambs are tightly tucked into the saddlebags. Then the herding collective would use the same ancient paths called *nal* that their ancestors had used to cross the Aravalli Hills before spreading out on the fertile plains of Southern Rajasthan and beyond.

On a mellow winter morning in December, we finally had time to catch up with the group and the three of us drove eastwards on National Highway 76 from Udaipur to Chittorgarh to meet up with Dharmabhai's *dera*.

Dharmabhai had been in sporadic contact with Hanwant since we had met him and a couple of days ago, he had let us know over the phone that his *dera* was roaming around somewhere near Mangalvar Choraya, giving a specific meeting point and time. However, since yesterday his phone had no longer been reachable. A little bit ahead of the meeting time, we pulled into Mangalvar Choraya, which was basically a crossroad of two highways that was lined with rows of transient pushcarts selling attractively arranged vegetables. A good opportunity to shop for supplies for our dinner with the shepherds. 'The vegetables are excellent here, much better than at home,' mumbled Dailibai as we tossed a bag of carrots into the back of the jeep, and in her usual quest to educate me about local ways, she added: 'That's because people here use *deshi khat* (manure), not *angrezi khat* (chemical fertiliser) to fertilise their fields.'

Before I could probe her further on the issue of *deshi khat* versus *angrezi khat*, Dharmabhai and another man suddenly appeared in front of us, leaning on their bamboo herding sticks. Both of them looked resplendent, their red turbans and white clothes setting them apart from the crowd of vegetable sellers clad in non-descript shirts and pants. To set it all off, they wore matchbox-sized, golden pendants of horse-mounted deities on their chests, giving them the air of important personages.

Dharmabhai introduced the other man as Bhuraram and the second in command, who helped him meet and talk to landowners and to represent the group to any bureaucrats or outsiders. The two men squeezed into our jeep and Dharmabhai directed us onto roads that got progressively narrower and bumpier. Finally, the path ended at a large open field. In the middle of it, seven charpoys were laid out in a square. Each charpoy was piled high with bedding and blankets, and women and children were moving around, busy with chores. Rows of goat kids and sheep lambs were tied up with loops around their necks; some of them had escaped and were frolicking about. In the corner of the field, under the canopy of some acacia trees, a few donkeys were milling around.

Dharmabhai introduced me to his wife, Hanzabai. Tiny and tough, with a weather-lined face, she smiled and welcomed us, spreading out a camel hair rug next to the charpoy and making Dailibai and me sit on it, while Dharmabhai led Hanwant away to a separate 'men's place' at the edge of the field.

Hanzabai started making tea on a little stove made from a metal bowl, strategically placed in the wind shadow of the charpoy. She and Dailibai chatted, while I watched the activities going on around me: small children playing with the lambs, the women, ranging from very young to late middle-aged, tidying up and rearranging their household gear. It was immensely peaceful here. Although we were actually not very far from the National Highway and you could hear the faint din of traffic in the distance, it felt like being in a separate universe.

'Hanzabai is the *patelani*,' explained Dailibai, 'because her husband is the *patel*, in charge of all the decisions of where the *dera* is set up and where to move.'

'This is a *dholri*,' she continued, pointing at the *charpoi*. 'There is one *dholri* for every family, and in each *dholri* there must be at least one woman to cook the food and make roti, and wash clothes, and so on. In this *dera*, there are seven *dholri*, meaning seven families. Altogether, there are 28 people in this *dera*.

'Each family has about 250–300 sheep and a dozen or so goats, as well as some camels and donkeys for carrying the luggage. At the moment, the sheep flocks and the goat and camel herds are out grazing. They will come back soon.'

While we sipped our tea, Dharmabhai and Hanwant were setting up our small tent, although the Raika themselves always sleep out in the open, unlike other nomadic herders who carry movable habitations such as tents and gers (yurts).

Dusk set in. Just as the red and golden ball of the sun dipped below the western horizon, a tremble went through the previously tranquil lambs: the seven main sheep flocks were returning from their grazing rounds and converging on the *dera*. A choir of bleating and moaning voices erupted as ewes and their lambs were overcome by the urge to connect, cuddle and suckle. But when there are thousands of sheep, it's no easy matter for mothers and children to find each other in the crowd. The shepherds scurried around trying to match ewes with lambs. It took at least half an hour and, as the light faded away, they did this using the light of their torches.

By the time the sheep symphony had ebbed into a soft finale, Dharmabhai and his colleagues had got a bonfire going and we all huddled around it. 'We', that is the Raika men in their red turbans and heavy, colourfully embroidered woollen blankets, Hanwant and myself. Dailibai was with the women who were in the centre of the woolly mass of sheep and preparing dinner on their movable hearths fashioned from broken strips of iron.

Dharmabhai explained that because many of the fields were not yet cropped, the farmers were on high alert, fearing that the Raika herds might destroy the harvest. Constant attention was required to stop animals straying into fields with standing crops and so avoid any conflict, and it was his job to prevent and soothe possible clashes. As the season progressed, more and more fields would become available for grazing and the attitudes of the farmers would change – no longer nervous about their crops, they would be grateful if herds were penned on their land because they benefitted from the manure that accumulated overnight and would boost their yields in the next crop cycle. Many of them actively invited the Raika to stay with them and provided them with cash and food items in return. The easiest part of the year for the Raika was the hot season that starts after the Hindu festival of Holi because then there was no more danger of damaging crops.

However, other worries and dangers were permanent, notably the gangs of thieves that regularly came to steal their animals, especially in the night. They drove up on motorcycles, with one person piloting and another riding pillion, and scooped up sheep. Only a couple of nights ago, there had been an attack and three animals had been lifted. So, every night, the Raika men took turns to be on vigil, while the women and children slept in the middle of the encampment, surrounded by a protective ring of sheep.

The police, however, were not supportive or making any effort to pursue the gangs. One horror story followed the next: how Raika were attacked here and there, how dozens of their sheep got run over by a truck, how they had to stay away from certain villages, how areas were closed for grazing.

Sometime during the evening, we were joined by the landowner, who brought some flour and tea in acknowledgement of the manure that the sheep deposited on his field during the night. He explained how the manure helped him to reduce the amount of urea, a nitrogen-rich chemical fertiliser he was using, and significantly lowered his expenditure on fertilisers. He emphasised – and this was later repeated by many people we talked to – that the local soils cannot take chemical fertiliser well and turn barren after repeated applications. We discussed the comparative values of different types of manure and sheep manure easily rated the highest, with the strongest effect lasting for three years, while buffalo dung is only effective for one year.

When we woke up in the morning and peeked out of the tent, the field was almost empty. The main sheep flocks and most of the men had already left for their morning round. Only women, children, lambs and kids were around. Hanzabai and Dailibai were sitting next to a charpoy and, as soon as they saw me, they waved me over and handed me a small metal bowl of tea.

While I slurped the tea made with goat milk, Hanzabai was rolling out flatbread made from maize flour. 'The men and the sheep will come back for breakfast in a little while,' explained Dailibai, shooing off one of the goat kids that rambunctiously climbed and then jumped off the charpoy. Two tiny human kids were playing with some of the goat kids.

I enquired about the milk output from the flock. 'While the lambs are small, they drink all the milk,' explained Hanzabai. 'But when they are bigger, or have been sold, we milk the ewes,' she continued. 'They only give half a litre or less, but it adds up.'

'They churn some of the milk into butter. Delicious!' Dailibai added. 'And many village people en route are keen to buy sheep butter or ghee. It is good for the treatment of some diseases.'

Later, two Raika men from a nearby *dera* came to invite us to visit, so we drove over and found an almost identical scene and setup. Here there were 12 *dholris*, laid out in a square, in exactly the same way. Whichever *dera* we visited the situation was remarkably uniform. And they all had uniform problems: thieves in the night, accidents when having to move on highways, and a total lack of official support.

All the *deras* we encountered had come from Godwar and it seemed as if they had systematically divided between each other the 'rights' to the harvested fields in Mewar. They come back to the same places and fields each year and have long-standing relationships with landowners, which are nurtured from both sides.

Over the next couple of days, we figured out that there must be around 40 *deras* from a fairly small area in Godwar roaming around in the region. When we calculated what this meant in economic terms, we arrived at humongous figures. Each *dera* averaging 3,500 ewes meant an estimated total of 140,000 ewes. Getting an idea of how many lambs they produce was a bit difficult as the Raika were hesitant to come up with figures, saying that the situation varied a lot from year to year, depending on rainfall and disease outbreaks. Apparently, the losses are high. A rough estimate was that 60,000 lambs survived. Of these, the female lambs were kept in the flock as replacements, while the male ones were sold to traders for meat at the age of a couple of months. A lamb would raise 2,500 rupees, and multiplying this by 30,000, we arrived at a figure of 75,000,000 rupees (equivalent to well over £1 million, depending on the exchange rate) of income generated for the herders.

With average lambs of the local breed weighing about 11kg (24lb), this added up to around 330,000kg (730,000lb) of live weight and 165,000kg (364,000lb) of lamb meat as an output of the system. All this protein was produced virtually out of nothing, without using any non-renewable resources, such as fertiliser, tractor fuel or transportation of feed! By means of migration, the shepherds of a small area in Rajasthan managed to generate what amounted to a second harvest from the land and to produce an enormous amount of food that would not have been possible if they had just stayed put in their home villages.

The other benefits are more difficult to translate into monetary terms. The value of the organic manure is priceless in a way, as it cannot simply be replaced with chemical fertiliser. Several farmers that we interviewed stressed that firstly, manure was much more effective and, secondly, that chemical fertiliser was not sustainable for the soil and undermined its fertility in the long run. The third product is the sheep milk, to which is attributed various health benefits and therefore often requested by farmers and other people encountered while on the move. And, of course, at certain times of the year, when the lambs are older, sheep milk is an important part of the diet of the Raika.

---

What the Raika of the Godwar area practise is termed agro-pastoralism – herding integrated with crop cultivation on a large scale. It is prevalent throughout much of India and forms the backbone of her food production and food security. However, in the more arid, less fertile parts of India where crop cultivation is patchy and only viable in very few years, herding takes on a different guise, in which animals have freedom to roam around on their own without constant supervision.

In the far west of Rajasthan lies India's Jaisalmer district. It is a harsh collage in tints of beige and yellow – sand dunes, gravel beds and calcareous soil, camels in various shades of dark yellow to light brown, mud-brick houses or the buildings of rich traders beautified by façades of elaborately carved golden sandstone. There is no greenery here and, even on India's rainfall map, it is the only beige patch in a sea of green, highlighting it as the sole place in the country where rainfall averages less than 200mm (8in) per year.

Jaisalmer's stark and vast open landscape abuts Pakistan in the west. The border arbitrarily divides what was once a culturally homogeneous area, the Thar Desert, into separate political entities. For hundreds of years, Muslims and Hindus lived here together peacefully, and still do, in an atmosphere of tolerance imbued with a Sufi attitude. The area has been alternately ruled by Muslims and Hindus, and people often converted from one religion to the other in line with the prevailing powers.

Despite now being studded with wind turbines and solar energy plants, Jaisalmer retains a sense of magic to which Indian and foreign tourists flock for a whiff of desert romance. They ride camels across sand dunes, glamp in luxury tents, revel in Sufi music under the stars and huff up to the majestic Jaisalmer Fort which sits on a craggy outcrop and is surrounded by massive golden ramparts that make it visible from tens of kilometres away.

Much of the land may not see any rain for years. Even crops supremely adapted to dryland conditions, such as pearl millet and guar, are risky propositions and hit or miss, depending on where rain falls during the monsoon season. But despite these almost permanent drought conditions, people have been living here for centuries, and, surprisingly, they have been living well. At least that's what they believe, although various official poverty indicators devised by urban people may tell a different story. According to a resident source, there is no poverty in the villages that blend into the landscape, so they are hardly discernible from a distance. This is due to the area's amazing wealth of livestock – sheep, goats, cattle and camels – which utilise the sparse and dispersed local vegetation and provide an abundance of livestock products as well as a steady flow of cash.

Jaisalmer happens to be the district in Rajasthan with the highest number of camels and, several years ago, this led us to implement a project there in support of the local camel breeders by strengthening them organisationally, providing camel healthcare services and marketing camel milk. One of our most knowledgeable and articulate partners was Mool Singh from Khabha, a small village about 35km (22 miles) west of Jaisalmer. Khabha consists only of a handful of houses spread out below a small fortress with a crenellated wall that once guarded an entrepot on the Silk Road. This earlier place is now in ruins, but from the generous floor plans of the collapsed houses it is evident that people were affluent. At some stage, Old Khabha was abandoned, and the place came to life again only when Hindu Rajput people from Pakistan moved here with their livestock after the 1965 Indo-Pakistan War.

Mool Singh is a stocky man in his fifties, wearing loose, white clothes and a tightly wrapped, multicoloured turban that accentuates his greying temples. Driving up to Khabha, we find him busy with a huge herd of camels that belongs to the village and has come for its daily visit to the watering pond. This is also the time when the camels are checked for injuries and treated. Four herdsmen are singling out camels that need treatment in the form of an injection or a wound dressing, a process that they resist with loud wailing and protest, but from which there is no reprieve.

Once this is done and the herd and its caretakers walk back to the grazing area, Mool Singh invites us to his mud-brick home that blends into the landscape. Next to the entrance is a pile of dried cow dung patties and a few metres away is a round stone enclosure that holds a goat kindergarten. We sit down with him in a small ante room that serves to receive guests, male guests normally. Women are not to be encountered by male non-family members in traditional Rajput homes, although they are active in the inner sanctum and kitchen, closely monitoring what goes on by questioning the young male attendants that file in to make us comfortable: cotton mattresses that were stacked up in a corner are rolled out on string beds, covered with colourful sheets in local motifs and the pillows fluffed up invitingly, so we can stretch out and relax. At short intervals, they bring trays with glasses of water, followed by tea in small cups and *numkin* (a salty snack). Desert hospitality.

Once we are all settled comfortably, Mool Singh gives us a systematic overview of the whole livestock system. He explains that the Thar Desert is characterised by a variety of very hardy trees and shrubs with extensive underground root systems that enable them to survive without rainfall for years. Most of them have tiny leaves to minimise the loss of water. The most important tree is the *khejri* (*Prosopis cineraria*), a medium-sized tree with a crooked stump that has extremely nutritious and delicious pods, which can be collected even in the direst of droughts. *Khejri* and other trees and shrubs such as *jhal* (*Salvadora persica*) and *bordi* (*Zizyphus nummularia*) provide excellent feed for livestock throughout the year; in addition, during the monsoon, wherever rain happens to fall, patches of nutritious grasses such as *sevan* (*Lasiurus sindicus*) pop up rapidly.

Mool Singh breeds camels, cattle, sheep and goats and earlier supplied bullocks to a fixed clientele of farmers in Gujarat who needed them to work in their fields. We ask him to describe how the camels are managed:

'Winter is the breeding season when male camels go into rut and calves are born. During this time of the year, we keep the herds near the villages to make sure everything goes alright with breeding and to be able to help when needed. If a mother dies, we bottle-feed the baby or get another female camel to adopt it. The first few days after birth, mother and young are together, but soon the mothers go out grazing on their own while the calves are kept in an enclosure. They come back in the evening to nurse their babies and spend the night with them. While the mothers are out feeding, we interact and play with the babies, so they get used to us.

'By the time the weather gets hot around March, the babies are big enough to roam about with their mothers and because forage and water are scarce now, they move further away from the village.

Several mothers and young form small groups and we don't keep track of them. They are on their own.

'In July or August, whenever the first rain falls, all of us camel owners go out to collect the camels. This is necessary to prevent them from destroying any crops that somebody may have sown. We trace the camels by means of their footprints and we have experts who can identify individual camels on the basis of these. The important part is that this is a collective effort: we don't just search for our own camels but catch any camel we find. We then bring them to a sacred place, a mosque near Lathi, where we conduct a ceremony to celebrate the successful reunion with our camel herds. Camels are branded with village and family-specific brands that everybody knows, so there is no dispute about which camels belong to whom.'

'So what benefits do you get from camels?' I ask.

'Well, one big benefit is the milk which once used to be more abundant than water, before the government started to pipe in water to villages. Earlier on, during severe droughts, this often saved our lives. The second benefit is, of course, the male camels that we used to sell at the fair in Tilwara and to the Border Security Force (BSF), for very good prices. Now of course, the situation has changed and it's becoming difficult to sell male camels. The BSF no longer needs camels, nor do the farmers, as they all go for tractors,' he says contemplatively, not complainingly.

'What about your other livestock, how do you keep it?'

'Every type of livestock is different. Cows are easy. They go out in the morning to graze and come back in the evening. We keep the calves here, so their mothers come back anyway, but we also feed them with leftover food and rotis. Well, cows give us milk which is churned into butter and then clarified into ghee. From buttermaking we also get buttermilk. And we use buttermilk to cook the fruits and pods of our desert trees, like *khejri*, *kumtia* and *khair*. Until recently, there was a good market for the bullocks. I had a lot of clients to whom I would regularly sell them. But that, too, has changed...same as with the camels,' he muses.

'Sheep always need a leader, so somebody has to go out with them during the day. Our sheep here have fine wool, but mostly we sell them for meat. Traders come and take them away. There is a lot of demand for them.'

'Goats are adventurous and forage on their own, but it is better to supervise them, as they create all kinds of mischief. We use their milk for making tea and we sell the male kids locally for meat.'

As if on command, our attendants march in and place brass plates in front of us. Next come millet breads, with a thimble of ghee poured over them. 'Desert vegetables', which are actually the fruits of the trees that Moolji described earlier and cooked in buttermilk, are spooned on our plates. Delicious yellow curry made from buttermilk is ladled out. A small bowl of curd is added and to wash it all down there is a glass of buttermilk mixed with cumin and salt. Topped off with a bowl of camel-milk *kheer* (rice pudding) as a dessert. A desert meal, heavy on dairy products, that deeply satisfies our stomach and senses.

---

It is often implicitly assumed that the movements of pastoralists are determined by a lack of grazing and the need to go elsewhere in search of better pastures. The common belief is that pastoralists are pushed to nomadise because they have exhausted local resources. Sometimes this may be the case, but reality is much more complex. Pull factors are also at work and may even be stronger. The Raika are drawn like magnets by the availability of biomass vegetation that can feed their sheep. The idea that somewhere, anywhere, grass is growing or trees are standing with nutritious pods is a magical attraction, because in their minds it would be a shame if this biomass was not put to good use. Often, when I travelled with pastoralists for advocacy purposes in India and beyond, I observed how they always keep their eyes open for anything growing, constantly evaluating it as a potential source of feed for their animals. If they see a green lawn somewhere they will say: 'We should have brought our animals.' Every time I come back from a visit to Germany, they enquire how the rains have been there and subsequently muse how they could transport their flocks of sheep there, and – in all seriousness – ask if I could not organise an aeroplane. The Raika that accompanied me to Germany were aghast at all the grass that was growing on roadsides and being mowed by municipality workers with expensive machinery. What a waste – it could sustain so much livestock!

Saverio Krätli developed the notion of pull rather than push factors behind the movements of pastoralists while studying the WoDaaBe pastoralists in Niger, a subgroup of the Fulani ethnic group that ranges all over Western Africa. The WoDaaBe are specialised cattle keepers, breeders of the famous mahogany-coloured Bororo breed which has long, lyre-shaped horns. They loosely subscribe to Islam but believe mostly in spirits that live in trees and whom they try to keep happy by following certain taboos in their interaction with the environment. They are also famous for their dance rituals in which the beauty of heavily made-up men is judged by women. Between 2002 and 2005, Saverio spent a total of 18 months with several WoDaaBe families in central Niger, joining them on their migrations and making detailed observations of cattle behaviour, herd management and breeding practices. In his PhD thesis he challenges many of the common assumptions about pastoralism. He concluded that for the WoDaabe, 'mobility is less a coping strategy than a pro-active endeavour to seek out qualitatively good grazing.' He made many observations that had eluded previous researchers – for instance, how the WoDaaBe support their herds to feed selectively on plants with high nutritional value at the right time, 'surfing the green wave' like wild herbivores. He also drew attention to the importance of learning in herding systems: young animals learn from their seniors which plants to eat and which to avoid. Saverio depicts the WoDaaBe herding system as a 'sophisticatedly managed interaction between cattle, forage plants, and the weather, and of an interplay between learnt behaviour and genetic predisposition.'[6]

According to Saverio and a group of British researchers, pastoralism actually capitalises on variability. It is flexible, makes best use of opportunities and copes with hardships one way or the other. It's a dance with the unpredictability that is the core characteristic of the system. In good years with ample rainfall, herds luxuriate in the lush vegetation, quickly putting on layers of fat that will see them through leaner times, and they reproduce bountifully. In periods of drought, they migrate to more abundant climes or lie low until it's over. The animals kept by pastoralists have the capacity to slow down their metabolic rate and thereby reduce their energy expenditure. In essence, they go into 'estivation', a kind of dormancy and the equivalent of hibernation for the summer. Reproduction is curtailed, pregnancy rates reduced, and newborn animals have less chance of survival.

High-yielding animals do not have this facility of slowing down their metabolism. They are continuously in high gear and cannot cope with an intermittent feed supply or exacting environments. If exposed to such circumstances, they just wither away.

The opportunistic, flexible and constantly adapting livestock production of pastoralists is anathema to scientific animal production with its emphasis on 'stability and uniformity', in which an artificially stable environment is created and everything is planned and standardised, from the genetics to the feed (precisely calculated) to the amount of (or lack of) space that is allotted for each animal, the number of days it takes to be ready for slaughter, the minimum amount of milk to be given or to face culling.

The difference between the two approaches in terms of resource prudence is phenomenal. Pastoralism is in a long-term equilibrium with the local availability of forage; it will never exceed it significantly or for long. If feed resources are decreasing, animal numbers will adjust downwards accordingly, actively supported by the herder who will, for instance, defer his female animals from becoming pregnant. This holds true as long as herds are sustained on natural forage throughout the year. It is when they are fed extra during summers or droughts that the equilibrium is destroyed and overgrazing occurs.

Herders take what I call a 'landscape approach' that seeks to optimally utilise the biomass in a given area, instead of maximising production. In order to do so, and not miss out on any nooks and crannies, they need animals with legs and brains that can climb up steep hillsides, swim through lakes, walk for miles between taking bites, and move far from water sources to track down edible forage.

In contrast, scientific livestock production centres on the animal itself in a 'portrait approach', in isolation from its environmental context. The output is impressive. But it externalises the associated environmental costs, such as for feed production by means of extensive monocrops of maize and felling rainforest to grow soya beans. It does not take into account negative impacts on water quality and biodiversity. The true cost of animal-sourced food is not calculated. The difference between the pastoralist approach and the 'scientific' method is that the former considers what is known as 'planetary boundaries', whereas the latter totally ignores these.

There is, of course, a range of intermediate options between the two extremes of pastoralism on the one hand and industrial production on the other, such as ranching and integrated farms. In ranching, animals are raised on extensive pastures that are fenced and privately owned. It is widely promoted in Africa as a progressive approach to livestock raising. But research in several African countries shows that the efficiency of ranching cannot compete with that of pastoralism.[7] Between the 1970s and 1990s, a large number of independent studies discovered that ranching actually produces lower yields per hectare than pastoralism. Notably in Uganda, pastoralism's returns were more than 6 times higher per acre than on ranches.

If we think about this, it is not really surprising, as herding allows attention to individual animals and better targeted grazing, and it requires fewer infrastructures. Scientists classify it as an 'extensive system', but it is actually extremely intensive with regards to both labour and knowledge. It has been fittingly described as the 'hardest work in the world'.[8] Certainly it is a labour of love for animals, but it is also a nature-based solution to our environmental woes.

# Nourishment

*Only if populations of animals and humans are spread out over the land will we be able to survive.*

GENE LOGSDON[1]

The most abundant – and most precious – output of herded animals is not meat or milk, but manure. An animal provides meat once in its life (or rather after its life), while only adult female animals produce milk, and that only for part of the time. But all members of a herd provide manure several times a day.

This manure is the lubricant of life, the link between animals and plants. It returns to the soil the nutrients that plants need to build leaves and stems, such as nitrogen, phosphorous, potassium and trace elements. It provides organic matter that aerates the soil and improves its ability to absorb water and hold moisture. Urine is even more valuable; it contains nitrogen in a form that can be easily absorbed and immediately made use of by the plant to synthesise the amino acids that are the building blocks of protein. Animal excrements – in the right dose – are a cog in the perpetual merry-go-round of nutrient flows, a component of the eternal earth rhythm that enables life.

The astuteness with which this cycle is managed determines the longevity of civilisations. Many of them came to an end because they depleted soils.[2] Others, in China, Japan and Korea, developed the knack of conserving soil fertility through an array of diverse strategies that included crop rotations, mixed cropping, leguminous crops and, most importantly, the right balance between animals and plants. These cultures considered manure, including that of humans, as valuable as a precious gem.[3]

---

The importance of dung sunk into me only when I visited the heart of India, the Deccan Plateau. Except in its river valleys, the Deccan is a vast and rather dry expanse where only the most drought-resistant crops, such as various types of millets, do well. In order to keep these crops going and growing, people co-opted sheep. The Deccan is sheep country.

The shepherds are variously called Dhangar (in Maharashtra), Kuruba (in Karnataka) and Kurumba (in Andhra Pradesh). Although they have different names, their cultures are very similar. They migrate seasonally or throughout the year with their flocks, deploying ponies or donkeys to transport their belongings.

I travelled to the Deccan, not to find out about manure, but because some friends of mine feared that the traditional Deccani sheep was gradually becoming extinct. This is a breed that is famous for being able to cope with drought and scarcity of fodder, that can walk over large distances and is impervious to extreme temperatures due to the special qualities of its wool, of which it produces large amounts. The shepherds weave this black and coarse fibre into blankets which form their signature attire and are important for ceremonial purposes. During colonial times and for a few decades afterwards, the black wool was much sought-after to make blankets for the Indian army.

But this market has fizzled out and presently there is hardly any demand for coarse wool in India and worldwide, as consumers opt for synthetics and very fine wool instead. Shearing is hard, time-consuming work and it's expensive; as a result, the wool of the black Deccani sheep has turned into more of a liability than an asset, making sheep breeds with hair rather than wool more competitive.

In order to kickstart the conservation of the Deccani breed, my friends were thinking of developing a Biocultural Community

Protocol (BCP). This is an officially recognised tool and document under the UN Convention for Biological Diversity in which communities describe their genetic resources and traditional knowledge and also identify the threats to them. Since I had gained some experience in the process from developing a BCP with the camel herders in Rajasthan, they invited me for a visit.

My two friends waited for me at the new airport in Goa. Nilkanth Mama[4] is a leader of the Kuruba shepherd community and his signature bright yellow turban made him stand out from the crowds waiting at the arrival gate. Gopikrishna is a social worker who helps a myriad of rural craft groups market their products to high-end urban customers. Both are involved in the struggle to conserve at least part of Karnataka's grasslands for livestock and prevent their takeover for yet another airport or factory. While our car wound its way through the steep escarpment of the Western Ghat towards Belgaum, Mama and Gopi filled me in on the history of the Kuruba and of the Deccani sheep.

They explained that, according to local belief, the Deccani sheep were originally made by Lord Shiva because he needed a *kambli* (blanket). But soon there were too many of them and when they created a nuisance, Shiva's consort Parvati put them into a termite hill. It so happened that two brothers were clearing and cutting down the forest to make room for agriculture. In the process, they also burned the termite hill and all the sheep came jumping out. Most of them had turned black because of being burned, although some remained white. One of the two brothers then started herding and became the forefather of the Kuruba.

Our first field visit was arranged by veterinarian Dr Bala Athani from the NGO Future Greens. Dr Athani quit government service to devote himself to rural development. He is going about his task systematically with an almost entirely female team that has set up thousands of female self-help groups who save money, so they can lend this among themselves. Leaving early in the morning, we were accompanied by his staff all clad in saris, which I knew from experience would still look immaculate and freshly ironed even after a whole day of bumping around remote villages over rough roads.

We encountered the first group of eight shepherds on a field not far from Bagalkot in a lovely and peaceful setting complete with lambs darting around and chickens from a nearby farm pecking away, and a number of hunting dogs and a couple of puppies frolicking around in the early morning sun. Women were feeding lambs with cut green branches and putting them into small enclosures where they would stay during the day while their mothers foraged on a nearby hill covered in scrub vegetation.

Dr Athani explained that the shepherds were Valmykis, a tribe whose traditional occupation was hunting. They had entered sheep breeding fairly recently, incentivised by the good income it provided. Their flock sizes were still small – averaging only 30 ewes per family. Even at first glance, it was obvious that the sheep did not belong to the black Deccani breed. Leggy and slender, they were white – some of them with freckles here and there – and, instead of wool, they had a sleek and glossy coat. Completely different to the black Deccani sheep that I had expected to be here. The shepherds seated themselves around us on the ground, forming a circle, and we started talking, but suddenly the owner of the land, Mr Reddy, burst onto the scene. Almost immediately he started gushing on about the value of the shepherds for his farm:

'Each sheep produces about 500g of manure per day, but I especially value the urine which is even more powerful. One hundred animals can fertilise my 2.5 acres in about 6 days. Having these sheep available means I can totally avoid purchasing chemical fertiliser. Therefore, I am giving them 6kg of *bajra* [pearl millet] per day for 100 animals.'

Since we had come to find out more about the Deccani sheep and not about farmer issues, we – somewhat unwisely in retrospective – tried to move the conversation back to the sheep themselves.

'Why don't you raise the black Deccani sheep,' I asked.

'We once had black sheep, but about 10–15 years ago, we became aware of this Yellaga breed. It grows faster than the black sheep, so we bought some Yellaga rams, and within a few years the breed changed.'

'We only keep the lambs for 3–3.5 months, then we sell them at the weekly market for 4,000 rupees,' another one explained.

'Yes, and there they are lapped up very fast to supply Chennai and Bangalore. There is almost unlimited demand and the price in the big cities is now Rs 185/kg of live weight.' explained Dr Athani.

He elaborated further: 'It's really a seller's market. So many new communities have started shepherding. Earlier it was only the Kurubas who were involved. But we did a study and it showed that 70 per cent of the shepherds in this area are newcomers. Members from both lower and higher castes have taken this up. On the other hand, quite a few of the Kuruba have shifted to farming and other activities. The money they made from the sheep allowed them to buy land or invest in other businesses.'

We learned many surprising things over the next few days. Shepherds used movable bamboo lattice fences in which they penned their sheep at night and that they dismantled and shifted every day to systematically fertilise fields one square at a time. In one village, rams were pampered with diets of eggs and treacle to participate in ram fighting, apparently an ancient sport in the area that dated back maybe a thousand years. There was the cult around Balamama, a simple shepherd who had come to be venerated not only by the Kuruba but by millions of others. His devotees herded flocks of thousands of black sheep on a volunteer basis. Farmers competed to host these sacred herds on their fields, which were accompanied by temples on wheels complete with sound systems airing religious songs. On each day of the trip, some new and astounding feature of the shepherding system was revealed, demonstrating how deeply embedded sheep culture was in the fabric of Deccani culture. The herders also worshipped wolves as saviours of their flocks because they performed the task of eliminating weak and sick animals.

With all these amazing ramifications to digest and make sense of, the economic importance of sheep manure only registered with me for good on the final evening when Mama invited us to his house for a farewell meal. Before serving dinner, he wanted to show us his own flock which was herded by his two sons. At dusk, we marched through a mosaic of potato and cabbage fields to the place where the sheep would be penned overnight. Tucked into the bend of a small, winding river, it was a harvested field on which about ten lambs had been confined in a covered little enclosure. We had barely reached the field when the flock appeared over the top of a small knoll, against the backdrop of the sinking sun. With two dogs that looked like German shepherds circling on the flanks, it was led by one of the sons with a red shawl wrapped turban-style around his head. Another son and a grandson in a baseball cap were following from the rear. Mama released the lambs from their coop and they raced towards their mothers with big hue and cry to suckle them energetically for a few minutes. The scene somehow reminded me of human mothers picking up their kids from nursery. During the reunion and as the sheep clamour gradually petered out, the men erected a portable enclosure made from bamboo poles and string netting. Nilkanth Mama explained that two of the crew would sleep out here overnight to guard, while the third would go home to his family. The owner of the field was paying 1 rupee per sheep per night for the penning. This was the regular rate in the area, although sometimes it could be as high as 2 rupees per head.

Later over dinner, which was served by Nilkanth Mama's daughters-in-law and granddaughters, we discussed the dynamics of the flock and worked out the economics. The herd was about 300 head, mostly composed of ewes that gave three lambs in two years. The main birthing season was November. About 15 per cent of the lambs were twins. Such a percentage was fine, said Nilkanth Mama, but if it increased it could spell trouble, as there would not be enough milk to go around and many of the lambs would die. For that reason, he did not think much of any attempts to increase the percentage of twins as promoted by scientists. In general, the mortality of the lambs was low and losses were rare, unless there was a disease outbreak. The lambs would be sold at the age of three to four months for around 4,000 rupees.

When we calculated the income, we concluded that the income from manuring or penning was actually even higher than from meat sales: 9,000 rupees – about US$130 in a month if all went well! This is a substantial sum in rural India, more than the monthly wage of a casual labourer.

Similar arrangements between farmers and pastoralists can be found all over India wherever crops are grown – for example, the manuring services of buffaloes and pigs in Odisha and of sheep and camels in Rajasthan. In South India, some cattle are kept predominantly for manuring and are never milked. In Tamil Nadu, semi-wild herds of Pulikulum cattle fertilise fields of rice, cotton and maize as well as peanut and coconut groves. The farmer's fee is calculated per hundred animals. While female calves are never sold, the male offspring is used for draught purposes and, in Jallikattu, a benign kind of bull-wrestling that is deeply embedded in Tamil and other south Indian rural cultures.

---

One of the explanations for the Hindu taboo on killing and eating cows is that even if they are unproductive, they still produce much-needed manure. Indian soils often do not respond well to chemical fertiliser, becoming hard and sterile after repeated applications. This seems to be typical for tropical soils.[5] The Hindu festival of Gobardhan celebrates cow dung, and at this occasion rural people decorate and worship a pile of cattle manure. And when Vandana Shiva, the famous seed advocate and legendary fighter against corporate control, was presented by her Monsanto antagonists with the Bullshit Award in the form of a pile of manure, she graciously accepted it as a tribute to organic farming and sustainable agriculture, noting that she felt honoured.[6]

In fact, farmers appreciate sheep, goat and camel dung even more than cow manure. The value of the manure and urine depends on the plants that the animals eat. Sheep and goat manure is rated higher than cow dung, as its effect is more intense and lasts longer. Farmers explained to me that this was because of the difference in feed: sheep and goats consume more roughage and fibrous material. Also understood is the fact that manure quickly loses much of its nitrogen content if it is moved around and exposed to the air. It is therefore crucial that it is integrated into the soil as quickly as possible after application and this is exactly what all the hooves of sheep and cattle do when the animals are penned closely together. No better system could be invented than this traditional method!

The doyen of organic agriculture, Sir Albert Howard (1873–1947), who conducted agricultural research in India for much of his life, considered the application of chemical fertiliser (which he called 'NPK mentality') as ultimately being the cause of crop, animal and human diseases. He was convinced that farming without animals was unsustainable and cited examples of large plantations of monocultures such as tea, coffee, cacao, sugar cane, rubber, coconuts, bananas, maize and cotton that gradually became unproductive until animals were brought back into the agricultural cycle.[7]

Surprisingly, the existence and value of the 'penning system' was never noted by animal scientists until very recently, although a couple of publications have now drawn attention to it, and it is even recommended by the World Wildlife Fund (WWF) for upholding soil fertility in cotton fields.[8]

A few years ago, Dr Athani and Gopikrishna presented a paper at the International Grasslands Conference in which they calculated the worth of India's manure at a massive US$45 billion a year, based on the market value of the nitrogen, phosphorus and potassium it contains. This not only has implications in terms of foreign currency which India saves by not having to import chemical fertiliser, but also with respect to the climate, as chemical fertiliser production is one of the most greenhouse gas-emitting activities there is.

---

If deposited on the land by migratory animals in discrete dosages, manure feeds and nourishes the soil; it is priceless. But in excess it becomes deadly. In modern livestock production it has morphed from a treasured resource into an extremely toxic substance because the balance between animals and plants and land has been lost. When animals are amassed on an industrial scale in what is known in the livestock industry as Confined Animal Feeding Operations (CAFOs), and feed is brought in from afar, manure piles up and its disposal presents a major problem. Pigs are not kept on bedding but on concrete floors, so their manure can be flushed out with water. This concentrate is then stored in open 'lagoons' next to the farms where it pollutes the air and regularly leaks into waterways. Eventually the 'nutrients' accumulate in lakes and coastal areas where the excess nitrogen and phosphorous spawn the excessive growth of phytoplankton and algae. This leads to 'eutrophication' – that is, the growth of algae causes a lack of oxygen and causes the death of aquatic organisms such as fish, crabs and oysters. One of the most feared manifestations is the 'red tide', a phenomenon that kills fish and damages marine biodiversity, including coral reefs.[9] It occurs frequently in south China and the coastal waters of the southern United States, causing casualties just like an oil spill. One of the worst examples for livestock production destroying the environment is seen in the hog factories in North Carolina. There the number of swine quadrupled between 1989 and 2003 – rising from 2.5 million to 10 million hogs. The excrements of the pigs are stored in huge open lagoons from where nitrogen evaporates into the air or poisons drinking water. Due to these factories – which have up to 850,000 pig inmates – an estimated 55,000 to 72,000 tons of ammonia end up in the air and the once ultra-clean rivers and coastal zones of North Carolina have become heavily polluted. In 1991, more than one billion fish died from poisoning. Nevertheless, in the late 1990s, the United States Environmental Protection Agency gave legal immunity to most of the nation's largest farms, which was supposed to last four years. This state of affairs continued until 2017 when the agency was rebuked and subsequently committed to take action, but again nothing happened.[10] In January 2021, President Biden promised to address the issue in the context of environmental racism (as it is minority groups who suffer most from the pollution).[11] At the time of writing, no further progress seems to have been made.

In October 2021, the Red Tide in Mar Menor in Murcia, Spain, made global headlines. This idyllic lagoon and top holiday spot that was once famous for its transparent water was covered in red slime and 4.5 tons of dead fish accumulated on the beach. This catastrophe has been attributed to the expansion of pig production in the vicinity and the runoff of manure. Due to heightened demand for pork from China (which had lost 40 per cent of its pig population because of an outbreak of African Swine Fever), the number of pigs raised in Spain increased by 3 million between 2019 and 2020. Nearly half the demand for Spanish chorizo, tenderloin and lard comes from China.[12] When Spain's agriculture minister made disparaging comments about factory farming in an interview with *The Guardian*, this led the agro-industry to accuse him of behaving like an activist and to calls for his resignation.[13] But fortunately, at least some limits are to be imposed on the number of animals that can be kept in such units.[14]

Similar collateral damage caused by too many animals concentrated in certain areas is now a regular occurrence around the world, including in regions such as Latin America that were earlier known for their extensive, ranching-based livestock systems. On 11 October 2021, an international coalition of 127 organisations from Argentina, Chile, Ecuador, Mexico and the US, many of them representing indigenous tribes, petitioned the Inter-American Commission on Human Rights to investigate factory farm abuses, making the case that the establishment of CAFOs is a human rights issue.[15]

# Creating Breed Diversity

*Why this diversity? The answer lies in the very diversity of the peoples and landscapes of Africa itself. The resulting diversity represents a living storehouse of precious genetic material that is becoming ever more important as modern methods of fighting animal diseases falter, the climate warms and markets change.*

Tadelle Dessie and Okeyo Mwai[1]

Academic textbooks state that animal breeding started in the late 1700s, when the progressive British farmer Robert Bakewell (1725–1795) discovered the benefits of selective breeding, as well as targeted inbreeding, and developed several new breeds of sheep and cattle. Another key event in the history of animal breeding is the start of the breed registry for the thoroughbred horse which started in 1791.[2]

This is another supremely Eurocentric version of events because, much before Robert Bakewell, the Bedouin – and probably many other pastoralists – had discovered the merits of selective breeding and it is their strict adherence to this principle and to purity that, for example, created the marvel of the Arabian horse, without which there would never have been the English thoroughbred. And half a millennium before Bakewell, there existed the Merino, a sheep breed that can be regarded as the ovine equivalent to the Arabian horse in its global popularity and widespread use in upgrading other breeds. Its extraordinarily fine wool enabled Spain to dominate the global textile trade of the time and generated the enormous wealth that funded its expeditions to the New World. The Spanish fiercely protected this asset from falling into the hands of anybody else and there was a death penalty on its export. But according to a very credible theory, they themselves had raided ancestors of the Merino from the North African tribe Banu Marin (Marinids) who briefly occupied the south of Spain in the twelfth century.[3]

---

The breeding strategies that created the Arabian horse placed prime value on pure breeding and pedigree. The most important breeding criterion of the Bedouin was not beauty or speed, but character: loyalty, intelligence and courage were desired. They traced pedigrees in the maternal line and were sticklers for pure breeding. Crossing pure 'asil' mares with non-pure-bred stallions was prohibited. Several travellers, including Swiss orientalist Johann Burckhardt and General Eugène Daumas, reported how they treated the horses as members of the family that had an open invitation to their tents if they needed protection from the elements. Mares routinely stepped into the tents to seek protection from midday heat or during a cold desert night. They were nourished with camel milk. And the Bedouin imprinted the foals on humans right after birth, by carrying newborn foals in their arms and walking them around the camp. For Bedouin children, the foals were playmates.[4] It was this combination of tightly controlled selective breeding and husbandry techniques that created the Arabian horse.

The Bedouin followed similar practices with respect to camel breeding. These have been documented in detail by Alois Musil, a Moravian linguist and ethnographer. In the first two decades of the twentieth century, he lived with the Rwala Bedouins who roamed the desert that has now been carved up between Syria, Jordan and northern Saudi Arabia. The Bedouin distinguished between thoroughbred and common camels. A thoroughbred camel had to be parented by a thoroughbred sire and a thoroughbred dam, and this had to be officially testified to by a witness. If no such witness was present, the offspring was not regarded as a thoroughbred because it could not be confirmed whether its father was of 'pure blood'. It was said that 'If a well-shaped female camel of the common sort is covered by a thoroughbred bull and the same is done with her female descendants, then a she-camel sprung from this line is recognised as a thoroughbred in the fifth, and the male camel in the ninth generation.'[5]

Keeping mental records of camels' ancestry was standard practice among other Bedouin tribes as well. The Al Murrah Bedouin traced genealogies through the female line for up to ten generations in the case of some of their most famous purebreds. The Harasiis of Oman memorise genealogical records and group camels (as well as goats) as 'daughters' of highly appreciated strains. When I interviewed the Rashaida Bedouin in Eastern Sudan who originated from Arabia, they could recite the forebears of their Anafi racing camels over seven generations, although they were less particular in this respect with their subsistence camels.

While Bedouin ways of breeding are the ones best known and most written about, most or all pastoralists follow similar principles, including Lake District shepherd and author James Rebanks, who elaborates on Twitter on how he needs to add another bloodline to his herd of Belted Galloway cattle and gets excited about how to select the right 'tup' for his family's flock of beloved Herdwick sheep.

---

Although the choice of the right stud animal is obviously crucial, pastoralist herds everywhere represent female power; they are overwhelmingly composed of female animals and their offspring. Females are the heart and core of a herd and structure it into blood lines that can be traced back to specific founder mothers. The Maasai conceive their cattle herds as being composed of 'houses' or matrilineages, and they give the same name to all animals descended from one particular cow.[6] The Raika do the same thing with their camels. Each of these maternal lines has a story about how its ancestor first came into the herd (often through a marriage, given as dowry or bride wealth). Each one of them comes with typical characteristics in terms of looks, personality, behavioural peculiarities, productivity and resilience. In herds from arid areas, there will be lines that are extremely drought resistant, but do not produce that much milk. There will be others that are high yielding, but succumb quickly in case of a climatic emergency.

For a herder, it is crucial to have the right mix. A good pastoralist strives for diversity in his or her herd to capitalise on the ability of animals to do well in favourable circumstances but that also have a good chance of weathering harsh years. Brigitte Kaufmann, a German researcher, describes this strategy in detail in her study of the breeding practices of different camel pastoralists in Northern Kenya.[7] She emphasises that pastoralists do not have the concept of an 'ideal animal', equivalent to the breed standard that exists in Western breed societies, and the same was noted for the Bedouin breeders of Arabian horses. Instead, they endeavour to own a portfolio of maternal lines that covers all eventualities. If there is an urgent need to gift or sell a female animal because of social obligations or cash requirements, it is chosen from a lineage that is well represented and not in danger of disappearing from the herd.

Whereas the female blood lines provide continuity, male stud animals come and go. For both biological and practical reasons, the percentage of adult male animals in a herd is usually tiny, no more than one or two per cent. Biologically, only one male is needed for impregnating between 50 and 100 female animals, depending on the species. And, in practice, it's a huge hassle to keep more than one adult male in a herd because, in typical macho fashion, they will engage in fights, maul and possibly kill each other in the struggle for dominance. It's the same situation as in the wild where male animals are wired to eliminate rivals. So, in order to have a modicum of peace, a herder keeps only the bull or stud camel that he thinks will father the right kind of offspring. The rest he or she sells or trades off – as a work animal, for meat or to another breeder.

The selection of the right male is obviously of paramount importance for the further development of the herd or flock. If the wrong specimen is selected, this can introduce unwanted genes for generations to come. Some herding communities go even further to ensure the right decision. Expert camel breeders, such as the Somalis, take into account appearance and behaviour, physical strength and the characteristics of the candidate's female ancestors (such as milk production, colour, resistance, etc.). Every year they single out two or three potential sires at birth, based on their pedigree, and provide them with special care and attention.[8] These are left uncastrated for as long as possible; as they mature they start to fight and the future stud bulls are chosen from the strongest and most dominant.[9] The Somalis even practise the equivalent of what is scientifically known as 'progeny testing', that is using a male animal more widely for breeding only after the quality of his offspring has become evident. They initially allow young potential herd sires to breed only a few females when they are five years old. If the offspring is good, the number is increased to 50 females when he is eight or nine years old.[10]

Most pastoralist societies have a ban or other restrictions on selling female animals to outsiders. They are seen as heritage that has been received from the ancestors and must be passed on to one's descendants.[11] Female breeding stock is not just individually owned but is also in the domain of the community. The rules for passing on animals to the next generation are often rigidly fixed, with animals transferred to children at specified life-cycle events, such as birth, circumcision and puberty, and as dowry or bride wealth in marriages. By not allowing any female animals to be sold to somebody outside the community, pastoralist societies in effect create closed gene pools, just like a Western breed society does.[12]

There is often a social obligation for more affluent community members to share their wealth with poorer families and clan members by giving animals on long-term loans. Such communal customs ensure that animals are distributed more evenly within the community – but only within – and that they remain a long-term community asset. This practice is not only a mechanism for more equity, it also helps to spread risk. For instance, if there are disease outbreaks, it helps ensure the survival of bloodlines. It's a custom that is of value even now. Among the Raika, if a man comes back from the city having decided that urban life is not for him, community members will each gift him a sheep, so he can reconstitute a flock and make a living from herding once more. (This was of major significance in the wake of the COVID-19 outbreak, when many young Raika lost their employment in the cities and returned to herding, often to discover that they enjoy being their own master rather more than being a menial worker in a metropolis.)

The sum of these pastoralist breeding techniques and social mechanisms very much mirrors herd-book breeding – except that the herd book is represented by the collective memory of the community, which is passed on orally, rather than being available as a paper printout, while purity is achieved by means of social mechanisms rather than herd-book by-laws. In essence, both approaches result in closed gene pools and thereby in distinct populations that deserve to be termed 'breeds'. Otherwise, there would be just one homogeneous gene pool.

A pastoralist herd is the result of countless management and breeding decisions over many generations. It represents the capital of a pastoralist family in a true sense. We can compare it to an assortment of stocks, some of which are high risk but promise high returns if circumstances are favourable (for instance, when rains are good) and others which will keep their value no matter what happens.

As an aside, the word 'live-stock' simply means a stock on hooves. Tellingly, the term and concept of 'capital' derives from the same root as cattle, stemming from the Latin word *caput*, meaning 'heads of cattle'. The concept of paying interest for borrowed money was taken from livestock that was given on loan, reproduced during the loan term, and then had to be returned with offspring. In many languages, livestock is etymologically linked to the words for wealth, and the term for money is derived from the word for livestock. For instance, *pecunia* (Latin for 'money') is based on *pecus*, a noun meaning 'cattle'. *Pecuaria* is the Spanish word for 'animal husbandry'. The

Sanskrit word *pashu*, meaning 'livestock', has the same root and is sometimes linked to *paise* which means 'money'. *Vieh*, the German word for livestock, goes back to the Saxon *fehu* that refers to livestock as well as money and is also the root of the English word *fee*.

---

There is a difference between human and natural selection. Nature selects for maximum aggressiveness among males, while a breeder is likely to look for other traits. Vitality is of course favoured, but aggressiveness towards humans is not, with preference for a friendly nature and ease of handling. Beauty, certain colours or colour patterns, milking ability and yield of female relatives, and wool quality are other criteria. These norms are not written in stone; they can change depending on market requirements and other external circumstances, such as climate change. A 'breed' is therefore never static, but always a work in progress, continuously being adapted as requirements change around it. It becomes static only once it is conserved in the form of frozen semen or eggs in a gene bank.

The Raika are very astute in responding to changing circumstances and adapting their breeding strategies to the needs of the hour. A few years ago, a Dutch student, Ellen Geerlings, studied the traditional knowledge around sheep raising in the Godwar area and documented how the Raika evaluated sheep and how they were making breeding decisions.[13] She reported that they were experimenting with introducing new breeds into their flocks that they came across during their migrations. One observation was that the traditional breed of the Godwar area – which the Raika called Boti – was becoming less popular. The Boti is small, with a white body, a black head and unusual tubular ears that look as if they have been docked, but are, in fact, natural. Another type of sheep of larger body size and with a red head and long pendulous ears was starting to predominate in the flocks. People said it was called Bhagli. Why was this happening?

Early one morning I wandered over to my shepherd neighbour, Jotaram, to find out more. He has a small, one-room house made from bricks and covered with handmade roof tiles. Adjacent to the house is a thorn enclosure filled with an estimated 80–100 sheep in various colour patterns: white and red, white and black, and tricoloured (white, black and red). Jotaram, a wiry man in his forties and as thin as a rail, was sitting on a charpoy, leaning with his back against the house and stuffing a clay pipe. He had just come back from his daily early morning excursion into the forest, where he had lopped a pile of fresh acacia branches that would serve as nibbles for the lambs during the day while he took their mothers out for their daily grazing round.

Jotaram told me to sit down and after an exchange of polite enquiries we got down to the question in hand. I said we had noticed that Boti sheep are becoming much less common in the area and that people are switching to the Bhagli breed with red heads. Why, I asked?

'Well, the Boti sheep was good for wool, it has very fine fibre. And it is also very drought resistant and tough. But nowadays, wool is not profitable. The costs of shearing are more than what we get for the wool. Actually, we have not been able to sell any wool in recent years and it is piling up in storerooms. A few years ago, Raika who go on migration to Madhya Pradesh brought back some ram lambs of the Bhagli breed from there. We saw that these lambs grow much faster and bigger than the Boti lambs ever did. So, we get a higher price for them from the traders. Also, the Bhagli sheep has longer legs, so it can walk further. It's much better for migration,' he mused, taking a series of puffs on his pipe and watching the smoke spiral up and disappear. Then he added thoughtfully: 'The Bhagli is not as tough. It cannot take as much pain and drought as the Boti, so I will keep some Boti in my flock. When there is a drought, I will use a Boti ram because the lambs are tougher.'

---

The capacity to thrive in specific eco-zones is not only a matter of genes, but also of training and acclimatisation. In Northern Kenya, the Somali and Rendille differ in their camel husbandry systems, resulting in quite different types of camels. The Somalis try to create ideal conditions for their camels, providing them with optimal access to pasture and ensuring frequent watering opportunities. The neighbouring Rendille, on the other hand, attempt to harden their camels by systematically restricting their water intake from an early age; they also allow them less time to graze. The resulting breeds are quite distinct – the Somali camels are large, heavy and good milk producers, whereas the Rendille camels are small and extremely tough.[14]

Pastoralists do not care as much about the individual animal as about their whole herd. What they need above all is a functioning group, an assemblage of animals that gets along with each other and is also obedient and respectful to their herder. So social aspects and ease of handling are very important for them. These aspects are not only determined by genetics, but are also a matter of inter-generational learning. In a herd, young animals learn from their mothers to trust and obey their owners and to get along with the other animals. In particular, they learn their entire feeding behaviour and knowledge of which plants to forage on. For example, camels cannot just be transferred from one type of environment to the other because then they are at a loss as to which trees to browse from. All over the uplands of Britain and Wales exists the practice of 'hefting', which makes use of the fact that female sheep are territorial and have a home range to which they are attached. Once they are 'hefted' to a particular area, there is no longer a need for fencing and they pass on this behaviour to their offspring, along with knowledge of the good grazing spots. This practice is especially important for sustainable grazing on the Commons, a large proportion of which is found in National Parks such as the Lake District. A hefted flock will automatically pass this knowledge from one generation to the next, without the herder having to do much.

---

We associate the human reign with biodiversity loss, and looking at the present avatar of our species, that is undeniably true. But herders have actually created biodiversity, as have farmers. Ever since humans entered domestic alliances with animals, they diversified them into what is known as 'domestic animal diversity'. The collaboration over the last 10,000 years with just twenty originally wild species resulted in 8,800 distinct livestock breeds, according to the Domestic Animal Diversity Information System (DAD-IS), a database that is managed by the Food and Agriculture Organization of the United Nations (FAO) and into which countries feed their data. Diversification has been especially high for sheep and cattle: there are 1,385 and 1,196 entries, respectively.

Domestic Animal Diversity is commonly measured by the number of breeds. But what is a 'breed'? The definition and use of this term has been somewhat problematic, as it is a Western concept associated with written records – that is, herd or stud books in which all animals that belong to a breed are registered. And they can only be registered if their parents are already on record. Furthermore, there is a breed standard to comply with, so animals need to have a certain colour or height or ear shape or tail form.[15] Thus a breed is commonly defined as a 'phenotypically distinct group of animals within a species' or 'animals that, through selection and breeding, have come to resemble one another and pass these traits uniformly to their offspring'.[16]

Application of this concept to the livestock populations in non-Western countries is often problematic. 'Especially in Africa, livestock breeds...are known by the same name in different locations, but often look quite different from one place to another. Conversely, there are breeds that look alike but have different names in different places,' stated a research report by the International Livestock Research Institute (ILRI), which is based in Africa and has made a huge effort to develop a database of the continent's animal genetic resources.

Other breed definitions give authority to breeders to decide what is a breed and which animals belong to it:

> A breed is a group of domestic animals, termed such by common consent of the breeders...a term which arose among breeders of livestock, created, one might say,

> for their own use, and no one is warranted in assigning to this word a scientific definition and in calling the breeders wrong when they deviate from the formulated definition. It is their word and the breeders' common usage is what we must accept as the correct definition.[17]

My own definition, which combines both social and ecological criteria is the following:

> A domestic animal population may be regarded as a breed if the animals fulfil the criteria of (i) being used for the same purpose, (ii) sharing a common habitat/distribution area, (iii) largely representing a closed gene pool and (iv) being regarded as distinct by their breeders.[18]

Breeds are referred to as 'animal genetic resources' by scientists and by the FAO, which leveraged a Global Plan of Action for Animal Genetic Resources in 2007, alarmed by the fact that around a fifth of the registered breeds were threatened by extinction. The process of breed diversification, which was earlier sustained by the fact that domesticated animals were subjected to diverse cultural regimes in an infinite variety of environments, is now in reverse, entailing a loss of domestic animal diversity that is estimated at two breeds per week by the FAO. This is a very dangerous trend, diminishing humanity's capacity to produce food in challenging environments and to adapt to climate change. Therefore, almost all nations signed the Global Plan of Action for Animal Genetic Resources agreed upon at the International Conference on Animal Genetic Resources held in Interlaken, in Switzerland, in 2007. Unfortunately, most efforts towards conservation have been focused on documenting and conserving breeds ex-situ, meaning on government farms or as frozen material.

However, this does not really serve the purpose. Saverio Krätli has been especially scathing in his objection to the use of the term 'genetic resources', pointing out that the functionality of animals in their respective ecosystems is only partly a question of genes and mostly of learned behaviour that is passed from one generation of animals to the next. Pastoralist herds are living heritage that, once lost, cannot be reconstituted like a *deus ex machina*. When hefted herds were killed off because of Foot and Mouth Disease, it proved very time-consuming and expensive to establish new ones.

There is no doubt that cryo-conservation is useful as a backup system for emergencies, but it would make so much more sense to apply resources and support the living systems. As gene banks require a significant amount of energy and infrastructure over an indefinite time period, it is expensive. Secondly, once genetic material is frozen, it is frozen in time. It no longer evolves in response to environmental challenges, for instance, with respect to disease resistance. And, as new diseases emerge, once unfrozen and re-constituted, the animals will be very vulnerable. Thirdly, there is the question of the foraging knowledge that young animals normally learn from their mothers and young pastoralists learn from their parents and grandparents – knowledge that can't be obtained from books. Without older animals to teach them and humans to give a guiding hand, a herd reconstituted from frozen material would be at a total loss.

Realistically, the only practical means of conserving diversity among our livestock breeds is to continue to maintain them in diverse natural ecologies. This in turn cannot be done without people with the necessary skills and dedication to survive in challenging environments. Thus, we need to conserve herding systems rather than focus only on genes. Simply seeking to conserve genes is another example of the reductionist approach that ignores the fact that biological diversity is a complex phenomenon with both a genetic and a cultural dimension. We depend on pastoralists to raise and steward the resilient breeds that will be of immense value in the future when we need to adapt to higher temperatures and more unpredictable weather patterns. They are the guardians of the livestock diversity that is essential for humanity to produce food in a solar-powered way that can heal the wounded Earth.

# PART TWO

# Herding Therapies

# Stewarding Biological Diversity

*If you have pastoralists, you do not need a national park.*
Jesús Garzón-Heydt, *Concejo de la Mesta*[1]

It's a mellow winter evening, and the ramparts of the Kumbhalgarh Fort that perches above our Camel Conservation Centre glow rose-red in the last rays of the setting sun. At this time of day, a succession of sheep and goat flocks led by Raika herders pass by our gate. The staccato patter of their hooves on the asphalt road indicates how anxious the ewes and nanny goats are to return to their lambs and kids waiting for them in the Raika hamlet down the road. But today, one of the herders, Tola Ram Raika, wearing a signature red turban and carrying a long bamboo pole on his shoulder, signals his herd to slow down and shouts to me over the gate: 'I need some medicine for my goat! She was attacked by a leopard.' He points at a speckled she-goat with a swollen neck and, on closer inspection, I can see small but deep bite wounds. We gladly provide some antimicrobial powder and ointment from our small dispensary. He continues on his way home, although the prognosis for leopard bites is usually very poor and he knows it. Following in his tracks a few minutes later is another shepherd, Devaram. 'For the leopard, it's not a sin to eat a goat,' he comments cheerfully and then adds: 'God gave it to the leopard in writing that he has the right to eat dogs and goats.'

Over the next few days, several more cases of leopard attacks are brought to our attention. But none of the Raika complains or is unduly upset, taking such losses in their stride as part of the natural course of things. And despite the illegal grazing, or maybe because of it, the conservation goals of the sanctuary with respect to predators are being met: the leopard population is growing steadily.[2]

On the Deccan Plateau in Central India, the Kuruba shepherds worship wolves because they believe they are necessary to keep their flocks healthy by eliminating sick and weak animals and thereby protecting the strong ones. For them it is natural selection at work, a process that also keeps sheep on the alert. The lazy individuals are taken; the ones who sprint away have better muscles and therefore fetch higher meat prices.[3]

When I visited a number of Kuruba shepherds in the sugar-cane-growing area of Karnataka a couple of years ago, they were deeply concerned about the wolf disappearing from the area. On every new moon, they explained, they worship not only their Goddess and the sheep pen, but also the wolf. When a wolf dies, they bury the dead body. And when an infectious disease hits, they leave a lamb in the wilderness to the wolves, believing that this will prevent further spread of the disease. Without the presence of wolves, they feel they have less protection against epidemics.[4]

This positive relationship between livestock and large carnivores is not only based on anecdotes told by and about herders, but it has also been confirmed by Indian wildlife scientists who established that the presence of cattle supports leopards, while sheep and goats are linked to wolves and hyenas.[5] The generally mutually supportive relationship in India between pastoralists and wildlife is not, however, officially acknowledged or taken advantage of. Instead, it is framed as 'human-wildlife conflict'. As a rule, pastoralists in India are not allowed to forage in protected areas, even though their elimination often has undesired consequences. In the Bharatpur Bird Sanctuary in eastern Rajasthan, a ban on grazing by buffaloes led to the disappearance of Siberian cranes. One of the theories to explain this is that the eviction of the buffaloes resulted in the proliferation of knotgrass (*Paspalum distichum*).[6] In their absence, the grass grew too high for the cranes to be able to spot approaching predators, so they no longer dared to nest.[7]

Not all herders have such a relaxed attitude towards predators as Indian pastoralists. In Europe, there is much resistance against the re-introduction of wolves, bears and other large animals in areas where they had disappeared. Battle lines have been drawn between herders and nature conservationists that promote rewilding, with each side having a different perspective. Many shepherds fear for their way of life, as the reintroduced carnivores prey on flocks and threaten to undermine the already meagre economic returns from shepherding.[8]

Globally, areas where pastoralists roam are often picked for setting up wildlife sanctuaries or national parks. This is no accident because if land is grazed, it means the original vegetation has not been replaced by a crop monoculture and the herd animals have acted as prey base for predators. There has also been no application of pesticides, herbicides, fungicides or other chemicals that kill soil life and the insects at the bottom of the food chain and those higher up that depend on them.[9] Biodiversity, therefore, has thrived.

'If you have pastoralists, you do not need a national park,' emphasises Jesús Garzón-Heydt, one of Spain's most respected conservationists. Garzón-Heydt, widely known as Suso, is one of the movers and shakers behind the revival of Spain's transhumance and its ancient drove roads, the *Cañadas*. Spain's network of drove roads goes back to the eighth century and was declared inviolable community property by the Castilian King Alfonso X in 1273 to support the semi-annual movement of Merino sheep between their winter quarters in the southern and coastal lowlands and summer pastures in the mountainous areas of the north. (For hundreds of years, the very fine wool of the Merino sheep was an incredibly valuable asset on which the Spanish Crown had a monopoly and so they would do everything to protect them.) The *Cañadas* were up to 75m (246ft) wide, had a total length of around 125,000km (78,000mi) and covered an area of more than 420,000 hectares (1,040,000 acres) of common property, equivalent to 1 per cent of Spain's total area. Until the middle of the nineteenth century, they were used for the regular movement of five million sheep, goats, pigs, cattle and horses. But they started falling into disrepair when it became possible to transport livestock by train at the end of the nineteenth and beginning of the early twentieth century, and later by truck. The availability of mechanised transport may have been convenient, but it had unexpected consequences for Spain's biodiversity: it led to overgrazing in the lowland pastures and the loss of the ecosystem services that the migratory herds provide en route between their winter and summer grazing areas.

I caught up with Suso in August 2021 when his herding group of Merino sheep, Retinta goats and ensemble of guard dogs was wandering around near Cervera de Pisuerga, in the mountains southwest of Santander, under the supervision of four herders. We arrived at the camp late at night after driving along a narrow winding road through forest that seemed impenetrably thick in the shine of the headlights. Quickly setting up small cub tents and going to sleep immediately, I had lost all orientation of where we were, but in the morning I was woken up by the baa of sheep, which were enclosed by an electrical fence that ran just a few metres away from my tent. Looking around, I found myself on an undulating ridge that separated densely forested land to the north from a steep decline into an agricultural valley to the south. Above the valley, but below me, hovered a cloud of mist that looked like cotton wool suspended in space. The glorious first rays of the rising sun quickly evaporated the wisps of white cotton, while the multi-species herding camp came to life. The herdsmen emerged from their tents and Suso from the car in which he had spent the night to go about the first order of business: reinforcing the ties within the team by feeding titbits to the Retinta goats – a threatened breed with long twisted horns whose main role is to act as bellwethers for the flock. They wore green collars, on which hung heavy bronze bells, and formed small crowds around the herders while the Merino sheep looked on. Many of them were stamped with the emblem of the *Concejo de la Mesta*, the herders' association that had historically administered the network of *Cañadas* until 1836 and which serves as inspiration to Suso. While the tall, dark maroon-coloured goats behaved affectionately, the huge guard dogs, reminiscent in size and skull shape of St Bernard dogs, but with a brindled colour pattern, kept their distance, apparently more bonded to the flock than the human party.

For fifty years, the now-75-year-old Suso has devoted his life to the revival of transhumance by speaking and writing about its benefits for nature, by raising funds, by supporting research and by providing practical logistical support to herders and encouraging them to stay in business. This year, he and his team accompanied 52 herding families, and since 1993 they have walked 122,930km (76,385 miles) of drovers' roads alongside transhumant families with sheep and other livestock – a protracted campaign that has disseminated 128 billion seeds and 77,000 tons of manure to date.[10]

Suso first realised the importance of transhumance for Spain's ecology when he was studying the *dehesas*, Spain's ancient wood pastures, in his native Extramadura, a landscape that consists of about 5 million hectares (12 million acres) covered by beautiful open forests of cork oaks, holm oaks and wild olives. This is also where 200,000 cranes and 40 million migratory birds from northern Europe hibernate and where the Spanish Imperial Eagle and the Monk Vulture survive. In this open area, millions of sheep, goats, cows and horses are grazed. This is where Suso started his work in 1966 at the age of 20. He was struck by the fact that almost all the trees were more than a hundred years old. Initially, he thought this was because agriculture had become mechanised and tractors had torn out young trees. But then he noticed that young trees were also absent in the steep and rocky locations where agriculture had never been intensified.

After decades of reflection, Suso came to the conclusion that the problem began when trains started running at the end of the nineteenth and beginning of the twentieth century. From then on, millions of animals from the Extremadura, Andalusia and New Castile no longer walked to the mountains, but were transported there by train from the lowlands, because the drought and heat in southern Spain cause the grass to dry out completely and there is no longer any water for them to drink. Earlier, millions of animals had to start hiking at the end of April to arrive in the mountains by mid-June when the snow had disappeared. This meant that during the whole of May and at least half of June the oak forests were free from grazing animals, which provided the opportunity for young trees to grow from the acorns that fall in October and November and then emerge above ground in April after they have developed a root system. But when the animals only leave in June instead of April, they eat the tender saplings. That is why it is necessary for the animals to walk from the lowlands to the mountains. As soon as Suso understood this, he resolved to revive the transhumance in order to protect the *dehesas* and all the animals that live there.

Suso began his project of reviving the transhumance when he was the director of the Montafragüe National Park. He used the opportunity to convince the EU's environmental office in Brussels of the value of the *dehesas*. Eventually they began to share his fascination for the *dehesas* and started providing funds for the revival of transhumance. Since this practice and the culture around it had already been lost for some fifty years, it seemed an impossible feat. But he somehow managed and, after a long-lasting and intensive campaign, the network of *Cañadas* was legally resuscitated in 1995 when Spain passed the Drovers' Roads Law (*Ley 3/1995 de Vías Pecuarias*), making it the first and only country worldwide where drove roads for the movement of herds are legally protected.

Suso's work has borne fruit and the pastureland buzzes with plants, butterflies and beetles. Vultures have made a comeback and Spain now hosts more than 90 per cent of the population of European vultures. The revival of transhumance ensures these birds of prey have a reliable source of food at certain times of the year.

The mosaic of different habitats that migratory animals create host a diversity of birds and insects, including pollinators. As Dr Paul Starrs, Professor of Geography at the University of Nevada, in the US, explains:

> In the mountains, the recuperation of pastures abandoned for so many years has regenerated diversity of plants, snails and spiders, beetles, grasshoppers, butterflies, kestrels, eagles, vultures, pheasants, quail, hares, deer species, and chamois.[11]

Research on Spain's transhumant sheep revealed that the *Cañadas* represent ecological corridors that facilitate the movements of plants and insects and that livestock act as 'seed taxis'. This phenomenon was first studied in depth by Pablo Manzano and Juan Malo who discovered that a flock of a thousand sheep can transport 200 million seeds during their 1,500km (900 mile)-long migration. In times of climate change, this is of phenomenal importance. It allows plants to move from a habitat they were originally adapted to but that has become too hot or dry, or otherwise hostile, into a new area that provides better conditions. Sheep act as a removals van for plants to an environment that is more to their liking, and this works not only for plants, but also for lizards, beetles and grasshoppers. Their service thereby prevents extinction due to climate change as it can be an important vehicle for plants to move into new areas that fit their requirements and thereby prevent their extinction.[12] In Germany, scientists have tried to put a price tag on the ecological services provided by sheep, who were found to redistribute up to 8,500 seeds from 57 species, and came up with the value of some 4,500 euros for a 200-head flock, if the seeds had to be purchased, transported and sown.[13]

Sheep don't just do this by default due to the nature of their wooliness; some herders make conscious efforts to disperse the seeds of preferred plants. An example for traditional regeneration is provided by pastoralists in the Islamic Republic of Iran who pack seeds in little bags and hang these around the necks of their sheep. During grazing the seeds drop through little holes in the bags and are worked into the ground by the sheep's hooves.[14]

It is not just that livestock transport seeds, they also aid their germination through scarification. 'Scarification' is a botanical term that refers to weakening the coat of a seed, so that it can break up and germinate. Many acacia trees, for instance, have very hard seed coats and their seeds need to pass through the stomach of a ruminant in order to sprout. The rescued camels that we kept in an enclosure on our campus for several months about five years ago left us a lovely gift: they have turned what was a bare field into a fledgling forest of young acacia (*Acacia leucophloea*) trees, a species that is otherwise difficult to propagate.

After they have scarified the seeds, livestock then conveniently trample them into the ground like a forest gardener. Trampling has other ecological effects as well, mostly positive, although it all depends on the intensity and the context. The depressions left by hooves can become filled with water and turn into mini habitats for insects and amphibians, which then provide food for all kinds of birds and mammals. And here we come to the general role of grazing animals at the bottom of the food chain. Their droppings are powerful incubators for a huge diversity of beetles and buzzing insects that not only feed populations of insectivorous birds, bats and reptiles, but also break down the manure into its constituents, which feed soil bacteria and loosen up the soil. The presence or absence of grazing animals in a landscape makes a huge difference to its biodiversity.

Grassland needs to be kept short and structurally diverse to provide suitable nesting habitats for many of our bird species that are under threat. Conservationists have therefore been mowing grasslands to create suitable conditions for them. But mechanical mowing is different from grazing and has had a disastrous effect on all small animals that live in grass, killing up to 80 per cent of insects, spiders, amphibians and reptiles. Some scientists correlate the massive decline in insect life that has worried ecologists and even the general public with the disappearance of grazing animals from the landscape. They regard the keeping of livestock indoors as one of the major factors responsible for the dramatic loss of insect, bird, amphibian and reptile populations that Germany has experienced.[15]

---

Restoring nature by permanently eliminating livestock certainly does not work. This lesson has been learned in many places. One example is the Zoigê plateau wetland on the northeast edge of the Qinghai-Tibetan Plateau. It is one of China's largest wetlands and, as the headland and watershed of the Yellow River Basin, it is of critical ecological importance for the huge downstream population in Southeast Asia. The government deemed it overgrazed, degraded and desertified, so the area was enclosed and reseeded. But in a small part of the restoration area, researchers allowed limited grazing by yaks and then compared various parameters that reflect soil health with the area in which grazing was not allowed. After six years of restoration, the grazed land had much better plant coverage, higher biodiversity, higher water content, excellent moisture capacity and was less acidic.[16] Similar observations have been made in other locations, including in the European Alps and in North America.

Grazing by domestic livestock has not only enriched biodiversity but also created some of our favourite landscapes. One of the most beautiful, biodiverse and species-rich habitats in Europe is 'calcareous grassland', which is rich in rare flowers – with up to 80 species per square metre, including orchids – and therefore represents a veritable paradise for butterflies. The origin of these grasslands can be traced back to the Neolithic Age when people first started to herd animals. They reached their maximum expansion between the fifteenth and twentieth centuries due to the presence of large, transhumant sheep flocks.[17] Thanks to ploughing, the use of chemical fertilisers and the disappearance of traditional shepherding, these fragrant, insect-buzzing habitats have become extremely fragmented and are seriously threatened – the largest remaining patch in the U.K. is across Salisbury Plain.[18] It covers 14,000 hectares (35,000 acres) and was preserved thanks to it being a major military training zone. Nevertheless, it became seriously degraded due to the lack of grazing among other factors, so in the early 2000s a restoration project was initiated with the goal of saving the Stone Curlew and the Marsh Fritillary butterfly. A critical component of the project was the employment of a herdsman and the development of a flexible grazing regime that avoided conflict with the army exercises. The approach was so successful that it was subsequently applied to other calcareous grasslands in Slovenia, Belgium and France.[19]

Pastoralists have created biodiversity not only in Europe, but also on other continents. In sub-Saharan Africa, for example, herders set up thorn-fence enclosures known as 'bomas' to keep their animals safe during the night. These enclosures may be kept in use for days, weeks or months. Due to the manure that has accumulated, abandoned bomas are nutrient hotspots that attract insects, birdlife and wild herbivores. This phenomenon is widespread in the savannahs of eastern, western and southern Africa. Recently, archaeologists determined that it goes back at least 3,000 years.[20]

The same pertains to Eurasia. Archaeologists working in the mountainous area of Kashmir and the flat lands of Kazakhstan conclude that pastoralists have shaped and eco-engineered landscapes as well as adapted to changing climatic conditions over millennia.[21] They suggest that we should restore the continent's grasslands not only by re-introducing wild species, but also by means of site-specific animal husbandry.[22]

Not for nothing have the German professional shepherds picked the slogan: 'Shepherds take care of the landscape that you love'. Pastoralism creates the kind of savannah ecosystems that our primeval ancestors were adapted to and which we still feel most comfortable in: open, so we can see a long way, but at the same time varied, so we can orient ourselves and have access to a variety of forage sources. Herds structure landscapes into a multitude of discrete micro-ecosystems by not moving evenly and mechanically, but seeking out their favourite plant types, which leads to uneven vegetation cover. Where they drop their manure, different plants start to grow. At the places where they rest at night, dung accumulates and attracts insects which then lure birds who leave seeds that grow into shrubs and trees that home bees. Each herded species has its own specific feeding habits and moves around differently.

In Europe, there is now a realisation that the original vegetation on much of the continent did not, as often believed, consist of dense forest cover but instead of a mosaic of grasslands, scrub and groves. This 'wood pasture' had co-evolved with large herbivores such as the European bison and the Przewalski horse, which are now extinct. In the absence of large, wild herbivores, grazing with domesticated animals is a necessity for restoring biodiversity. In Germany, grazing projects are the most frequently implemented by nature conservation measures and shepherds, who derive the majority of their income from contract grazing in nature conservation areas, with the goal of conserving specific plants and ecosystems which developed through interaction with large herds of wild ungulates.

Sadly, the huge herds of herbivores that once dominated the world's rangelands are gone. Hopefully not for ever, but in the meantime, we have the option of micro-managing ecosystems with the help of domesticated herd animals in a targeted way, so they can connect habitats and create patchy landscapes. If managed skillfully, herded animals can not only do that, but also help to maintain soil health and cool the planet by returning carbon to the ground.

# Regenerating Land

*The health of soil, plants, animals*
*and humans is one and indivisible.*

SIR ALBERT HOWARD[1]

'Regenerative grazing' is an emerging trend in both the US and U.K., heralded as a way of restoring soil health and biodiversity and addressing climate change as well as being a viable farm business. This ecological movement is rapidly gaining new followers and has adopted many of the principles and practices of pastoralists – with some innovations, such as managing the movements of animals by means of daily shifting electric fences rather than via a close human-animal relationship.

The pioneer of this movement is Allan Savory, a white Zimbabwean, who is now in his mid-eighties and originally worked as a wildlife conservationist for the government of what was then Northern Rhodesia. Worried about overgrazing, he ordered the culling of thousands of elephants whom he believed were destroying their habitat. He soon came to regret this as he noticed that land protected from grazing took a turn for the worse. He eventually developed the theory that herds of animals are necessary to uphold soil processes by depositing dung and trampling seeds into the ground. Unless animals eat the grass and trample the dead plant material into the soil, this accumulates instead of decaying, thereby blocking fresh growth, leading to a loss of soil carbon, the eventual depletion of the soil and its inability to sustain plant and animal life.

Madhuram Raika, our camel herder.

Madhuram is filling the cups of our neighbourhood children with their daily dose of camel milk.

The Raika women took to the microphone like born entertainers.

Raika women at the Global Gathering of Women Pastoralists in Mera.

Dried dung patties are a valuable source of energy for cooking and for which no cash has to change hands.

Early morning encounter of the author with friendly camels at the famous Pushkar Camel Fair. Image by Hanwant Singh Rathore.

Raika camels are avid kissers, but only if they are treated right and a relationship is built up from birth. Image by Ramesh Bhatnagar.

During migration, women lead the herding group.

After arriving at the campsite, the camels are unloaded and the camp is set up quickly, usually in the same formation, every night.

Ramming in the posts for the movable pen.

The sheep are enclosed in pens that are moved systematically across a field to manure the soil.

Nilkanth Mama is the leader of the Kuruba shepherds of Karnataka, in central India.

The relationship with the bellwether, which carries the insignia of the historic *Concejo de la Mesta*, must be continuously reinforced by feeding him daily titbits.

Jesús Garzón-Heydt (aka Suso) in his element surrounded by sheep, goats and dogs.

The Indian Globe Thistle (*Echinops echinatus*) is a weed for farmers, but camels love it and transform it into very sweet and very healthy milk.

Pastoralists advocating for Livestock Keepers' Rights during the Interlaken Conference held in Switzerland. Only the governments of the African region supported this. Image by Sanjay Barnela.

The Raika delegation enjoying a field trip during the First International Conference on Animal Genetic Resources held in Interlaken, Switzerland. Image by Sanjay Barnela.

Dailibai hitching a ride on an oxcart during the passing of the herds through the centre of Madrid.

The Raika were one group of many pastoralists joining the annual passage of transhumant sheep flocks through Madrid.

A flock of sheep crossing the emblematic Ariza bridge across the Guadalimar river in Úbeda (Jaén, Spain) that dates back to the Renaissance period in the 16th century. Image by Katy Gomez.

A sheep flock migrating across a dehesa in the Sierra Morena of Jaén, Spain. Image by Katy Gomez.

During the autumn migration, livestock crosses a stream of the Muso River in Jaén, Spain. Image by Katy Gomez.

The Kumbhalgarh Camel Dairy makes a point of keeping camels in mobile systems and does not support stall-feeding.

Recent research on the impact of large herbivores such as elephants and wild boars further supports Allan Savory's hypothesis. The role of herbivores in the ecological cycle is to reintegrate the carbon that has accumulated above ground back into the soil, so it can nurture soil micro-organisms and root growth. Without herbivores, the whole natural carbon cycle comes to a halt, and the vegetation amasses on the surface of the soil instead of being re-integrated, releases methane and is prone to fires. This is where soil health and climate change connect and why we urgently require herbivores in the landscape – to get carbon (and other nutrients) below the ground and into the soil! It is their trampling that gets organic matter into the soil, thereby enabling it to draw in more water and provide better conditions for seeds to germinate and for saplings to grow more successfully above and below ground.

On the same lines, the Russian scientist Sergey Zimov has developed the theory that the extermination of the Arctic megafauna by human hunters during the last Ice Age (Pleistocene) led to the disappearance of grasslands in Siberia and thereby the loss of the area's ability to absorb carbon. In the Pleistocene Park, located in northeastern Siberia, he is seeking to reintroduce large herbivores to recreate the grasslands and thereby counter global warming.

Based on his observations in Africa, Allan Savory developed the concept of Holistic Resource Management which is now followed by a large number of ranchers around the world. It involves grazing with a plan to keep moisture in the soil from the end of one rainy season to the beginning of the next. The Ted Talk in which he shared his credo entitled 'How to green the world's deserts and reverse climate change' has been watched more than four million times.[2] The details of his theory can be quibbled about and there is still much research and experimentation to be done on which type of livestock or which combination of herded animals does best in specific environments, but the basic principles are unquestioned: the world's grasslands have co-evolved with herds of ruminants and their continued existence requires on-going interaction with grazing animals to be maintained. Planned and carefully monitored grazing has the capacity to influence soil health and biodiversity; it can be a valuable tool in conserving certain plant and animal species.

The secret to regeneration grazing is the short-term use of pastures by 'mobs' of herd animals, alternated with long periods of rest, if we are to come close to the way wild herbivores used to move through the landscape on the world's steppes, savannahs and prairies. The idea is to fill the gap left by the wild herds that earlier roamed the grasslands of the world, such as the millions of bisons that once moved across the Great Plains of the US and Canada, creating rich fertile soils and capturing carbon below the ground.

How this is achieved in practice differs. In the rangelands of the Western US, herds are moved cowboy-style from horseback and the practitioners are dubbed 'carbon cowboys'. In the Eastern US, where rainfall is higher and land is scarcer, pastures are often subdivided into small, fenced plots and herds moved from one to another on a daily basis – a practice known as Adaptive Multi-Paddock Grazing (AMP). This has been shown to support higher livestock densities while resulting in higher soil carbon, denser vegetation and better water infiltration compared to continuous grazing.[3] However, as its promoters are emphasising, this is not a formula or recipe, but a system that requires close and continuous observation of animals and pastures so that it can be adapted to changing situations. Which is exactly what pastoralists are doing.

---

Regenerative grazing has adopted many pastoralist principles, such as keeping animals on the move and selecting for those that do well without added feed, have good mothering instincts, are docile and can look after themselves. Yet there are important differences between the herding cultures that are the focus of this book.

Firstly, regenerative grazing is a phenomenon largely performed in the temperate parts of the world where rainfall is fairly reliable and occurs throughout the year, whereas pastoralism operates in drylands that see only minimal and seasonal rainfall which fluctuates significantly between years. Secondly, ranchers, or the graziers that they hire, work on private or leased land and thus have secure access to their grazing land and can plan ahead, while pastoralists manage their animals on the Commons and/or on farmer-owned harvested fields and therefore can never be sure what to expect when they return after a year's absence.

While regenerative grazing is hailed as a progressive and ecological approach in the 'North', pastoralism in Asia and Africa is still widely stigmatised or ignored. Mention pastoralism to almost any policy maker, development planner or even many a biologist, and they will almost automatically associate it with overgrazing, drought and desertification. I have often heard so-called experts blaming pastoralists for creating deserts. By irrationally keeping too many animals, they say, herders have exceeded the land's 'carrying capacity' and transformed formerly fertile areas into 'wastelands'. Another concept often referred to in this context is the 'Tragedy of the Commons'. We need to put all three terms – desertification, carrying capacity and Tragedy of the Commons – under the microscope.

What is desertification? According to the United Nations Convention to Combat Desertification (UNCCD), desertification is the process of land degradation in arid, semi-arid and semi-humid areas, mainly caused by unsustainable land management and climate change. It goes along with the loss of productivity and fertility of drylands, leading to their abandonment.

When the three international environment conventions – the United Nations Convention on Biodiversity (CBD), the United Nations Framework Convention on Climate Change (UNFCC) and the UNCCD – were set up in 1994 as a result of the UN Environment Summit held in Rio in 1992, the UNCCD was intended to cater especially to the interests of African countries. The choice of the term 'to combat desertification' provides insights into how desertification and its control were conceived at the time. The wording conjures up images of militaristic operations that stop advancing sand dunes from encroaching upon and burying human civilisation.

But the concept of desertification is scientifically wobbly and mostly used as an instrument of political power, according to a group of highly respected scientists. In a volume entitled *The End of Desertification*, edited by rangeland scientist Roy Behnke, they note that the establishment of the UNCCD in 1994 was a reflex to a series of drought years that affected the Sahel from the 1960s to the 1980s. These were perceived to cause the Sahara Desert to expand southwards by 5–6km (3–4 miles) annually. However, since then climatologists have revised this assessment by using satellite remote sensing in combination with on-the-ground studies of the vegetation and concluded that such an expansion did not actually take place. Furthermore, they are clear that the droughts were not the result of human mismanagement of land and 'overgrazing' but instead due to weather fluctuations, over which humans had no control.[4]

The authors of this landmark book regard the concept of desertification – which implies desert encroachment on previously fertile land – as not useful and warn that many approaches to combating desertification are misguided. They doubt the UNCCD's rationale for combating desertification. Instead, they suggest, it's an issue that must be addressed through local actions. Desert communities throughout the world have done remarkably well in adapting to what others consider challenging environments and developed a diversity of social-environmental systems that have proven very resilient. We cannot 'combat desertification'. Instead, we have to love deserts, says Rajasthan's 'waterman' Rajendra Singh, founder of the NGO Tarun Bharat Sangh, who has managed to get rivers to flow in what was previously considered a desert – all by mobilising local people and reviving their traditional technologies to harvest scant rainfall and make the most of it.

The term desert derives from the Latin *deserere*, meaning 'to abandon, to leave, forsake'. But deserts are not at all abandoned. Instead, a huge number of plants and animals and people have made them their home. Much of this life may be dormant and only bursts onto the scene when rain happens to fall. Desert plants are experts in withdrawing underground, sometimes losing all their above-ground parts, but maintaining huge root systems that enable explosive growth as soon as there is even minimal precipitation. Desert animals are active mainly at night or have developed a panoply of physiological mechanisms that render them resilient to extreme heat and able to get by with minimal water consumption. People may lie low to save energy or seasonally withdraw into more fertile areas.

Dryland communities have developed a host of local technologies and created social institutions that enable them to live in 'deserts' quite well, with the help of drought-adapted livestock, especially camels. They have invented ways of capturing, storing and channelling rainwater – for instance, in Iran through extensive systems of underground aqueducts, known as *qanat*. In Northern Arabia, the Nabateans developed extensive water collection and storage systems that enabled them to grow crops.

Among the traditional institutions that pastoralists developed to survive in deserts and drylands are rules to protect areas from grazing and keep them on reserve for severe droughts. One example is the sacred groves (*oran*) of Rajasthan. These were under the guardianship of local deities and were places where the cutting of trees was prohibited. But they were thrown open for grazing during severe droughts. In the Arabian Peninsula, the Bedouin developed a system called *hima* that operated according to similar principles.[5]

Such community-developed and managed land management systems fly in the face of the 'Tragedy of the Commons', a concept which states that because nobody bears responsibility for the upkeep of resources that are not privately owned, they are doomed. This theory was elaborated by ecologist Garrett Hardin and published to great acclaim in 1968. He used the village commons to illustrate his point. Village commons are a long-standing institution, not only in England, but in many European countries as well as in India, and serve to provide grazing for village cows or other livestock. It works fine, Mr Hardin said, as long as the carrying capacity is not exceeded. But if only one farmer added a cow above the carrying capacity, this would profit him individually but lead to the degradation of the commons at the expense of the larger community. His implication was that the commons was a kind of lawless territory with no rules and nobody taking responsibility for its upkeep, which led to its overexploitation and degradation. Thankfully, the Nobel Prize-winning economist Elinor Ostrom rephrased the issue as 'The Drama of the Commons', documenting and drawing attention to the rule systems that many communities have developed to manage their commons sustainably.[6]

'Carrying capacity' is an ecological concept describing the maximum number of animals of a species that a given environment or area can sustain without degrading. This concept is widely used, especially by rangeland managers. Yet its applicability to drylands has been questioned by the same people who doubt the validity of the 'desertification' concept. Deserts are characterised not only by low average rainfall, but also by enormous annual fluctuations in precipitation – many years of low rainfall are interspersed with those in which it is quite high. They represent 'non-equilibrium rangelands', argues Roy Behnke based on his experience working with pastoralists in Africa, Mongolia and elsewhere. Drylands differ from 'equilibrium rangelands' in temperate areas where precipitation remains more stable from year to year and where the carrying capacity concept makes sense. Roy's theory is that 'non-equilibrium rangelands' do not have a fixed carrying capacity and are not subject to degradation – except around water holes and so-called key resources where livestock might accumulate in greater density and linger for longer. Instead, livestock numbers automatically adapt to the changing availability of pasture resources. In a bad year, or a series of them, when forage resources become scarce, the weaker animals die and those that survive do not become pregnant or their offspring succumbs. Populations drop. After a while there is a year with good precipitation leading to a higher reproduction rate and an increase in livestock numbers. There may even be more biomass available than the existing livestock can consume. But the next drought year will inevitably come again, adjusting livestock numbers downwards. So, population numbers are usually in long-term sync with the carrying capacity, which fluctuates from year to year. However, this equilibrium can also be disturbed if traditional knowledge has been lost.

In a nutshell, the beauty of pastoralism is that it is self-regulating. It never exceeds the carrying capacity of the land for long because nature automatically corrects this.[7] Only when livestock numbers are artificially inflated by providing animals with supplementary feed during part of the year is this balance disrupted, and this can lead to degradation. But even that is reversible if the root systems of forage plants have not been removed. By contrast, land use changes such as ploughing and replacing drought-adapted native vegetation with water-guzzling crops cause permanent damage. The Thar Desert is blessed with a variety of trees and shrubs that are excellent livestock forage, available throughout the year and also extremely drought resistant. The shrubs have extensive root systems, and although their above-ground parts may almost disappear during severe droughts, they come back in full force when rain falls. But in large areas, the trees have been uprooted and the land is irrigated from tubewells to grow mustard or wheat. Once the groundwater is finished, the land lies bare and uncovered, with winds blowing away the topsoil.

The important point is that the vegetation has time to recover from grazing, and this is what pastoralists try to ensure through their mobility and seasonal migration cycles. Hilde Gauthier-Pilters was an ecologist who studied the feeding behaviour of free-ranging camels in the Sahara.[8] Her conclusion was that camels are not damaging for several reasons, and that they actually trigger plant growth: herds spread out widely, so there is no bulk grazing. They take only one bite from each plant before wandering over to the next one, triggering the plant to grow more. Madhuram has mentioned to me repeatedly that camels do not like to go back to the trees they or other camels have browsed on, leaving them to rest for an extended period.

One example that demonstrates the importance of keeping livestock on the move is furnished by the experience of China with its changing policies for the nomads of the Tibetan Plateau. This vast expanse serves as a water tower for major Asian rivers, including the Mekong, the Yellow River, the Yangtze, the Brahmaputra and the Indus, on which billions of people downstream depend. Like a giant sponge, the grasslands soak up the melted snow from the mountains and then release the water slowly. Without grass, this function would be impaired and the water would cascade down torrentially rather than in a steady stream. So, its ecological integrity is crucial.

The Tibetan Plateau has long been home to a large number of nomads who traditionally moved their herds of yaks, sheep and goats between winter pastures in the lowlands and summer grazing areas in the highlands, living sustainably from their products. Considering this inefficient, the government started to allot land to individual nomadic households in the 1980s and forced them to fence their plots. The ensuing severe restriction of movement, necessitating continuous grazing in a small, confined area, led to degradation of the vegetation, soil erosion and a loss of animal productivity. Blaming this on nomadic ignorance and worried about the role of the Tibetan Plateau as a source of water, the government designated huge areas as national parks and removed around a million pastoralists. But this exclusion did not have the expected positive effect.[9] Instead, biodiversity was reduced, at least in the short term.

By now there is a consensus that the Tibetan rangelands are not a natural landscape, but largely anthropogenic. Scientists believe nomads have shaped these grasslands intentionally to ensure good conditions for their herds and that they have done this by means of the combined use of grazing and fire over centuries or even millennia. Fortunately, there are indications that the attitude of the government is slowly changing and controlled grazing is beginning to be used to reverse degradation.[10]

Overall, we can conclude that pastoralists have, or had, a variety of mechanisms to maintain the balance between their environment and the size of their herds. The problems start when people and governments from the outside come in and implement radical changes in land use patterns above the heads of the local people and ignore the wisdom and experience that they have accumulated over many generations.

---

Another controversial issue is the water footprint of livestock. Cattle, especially, are often routinely accused of being water guzzlers and we frequently read statements that it takes 2,040 litres (450 gallons) of water to produce a quarter-pound beef patty.

For herding systems, these indictments surely do not apply. The high-water footprints ascribed to meat, and that are unfavourably compared with those of cereals, are calculated on the assumption that animals are fed cultivated feed grown with irrigation and fertilisers. So, 98 per cent of the water footprint of cattle can be derived from growing feed. But if cattle just graze on pastures and natural vegetation, their water footprint is minimal and restricted to drinking water which makes up a mere 1.1 per cent of their total water requirements.[11] In temperate areas, pastured cattle obtain 70–90 per cent of their water requirements from grass. And much of the water they drink, they then return to the soil through their manure and urine.

In fact, livestock can produce food in water-deprived areas that do not receive sufficient rain to grow crops. We can again use the example of the Thar Desert where camels and goats convert the leaves of drought-adapted trees into milk and meat. Currently, water is channelled there from the Himalayas by means of the Indira Gandhi Canal; in addition, a multitude of tubewells has been sunk. But the elders still remember when water was rarer than milk and people obtained their fluid requirements from the milk of camels and cows who in turn sourced it from vegetation.

Camels are the species with the lowest water requirements. Their legendary ability to go for long periods without water is not just a question of physiological adaptation that allows them to minimise their water expenditure. It is also a matter of pastoralist management and, to some extent, a question of training and upbringing. The Rendille pastoralists in Kenya limit the water intake of young camels in the belief that too much water softens them, with the result that they can go for a week or longer without drinking even when little water can be obtained from forage. In the Horn of Africa, the general rule is that camels need to be watered every seven days, sheep and goats after three days and cattle daily. This determines the distance from which they can forage from their watering points.

To be sustainable and use resources in tune with their natural availability, we must ultimately go about producing food differently, and, in low rainfall areas, raise livestock instead of growing crops by means of irrigation. California is a case in point. The state is known for its agricultural output, but its own groundwater resources have been almost depleted due to decades of intensive agriculture in the Central Valley. Now much of the water is sourced from the Colorado River or is channelled from Northern California to Southern California. Water allotment in California is a hugely controversial issue; nevertheless, a large chunk of it, about 18 per cent, continues to be used for irrigation to grow the alfalfa that feeds the dairy cows of the state, but is also exported to places like China, Japan and Saudi Arabia to feed livestock there.

If one wants to produce food in Southern California without resorting to imported or pumped water, then goat raising is the only option, believes Gloria Putnam, an electronics engineer who is concerned about the conservation of resources and finding methods of sustainable food production. In an 'intellectual project', as she calls it, she bought 28 hectares (70 acres) of chaparral and woodland in the San Gabriel Mountains near Los Angeles, acquired a herd of goats that gradually grew to 50 female breeding goats, and set up the Angeles Crest Creamery.[12] Having no understanding of the forage value of the natural vegetation growing on her land, Gloria had to learn everything from the goats by herding and observing them closely. She learned that, just like Madhuram's camels, her goats vary their diets through the course of the year, seasonally focusing on specific plants. After grazing/browsing certain patches, they only return to them the following year, which gives them time to recover. In essence, they developed a kind of 'adaptive multi-paddock' grazing system on their own, without human guidance. The shrubs respond well to being browsed and come back denser the next year, although Gloria does interfere if she feels that the goats might overgraze, especially where the land is recovering from a bushfire a couple of years ago. Gloria believes that within her lifetime the only viable agricultural system in Southern California will be goat raising, as it becomes politically untenable to keep importing water from Northern California or other states. 'The only other alternative will be hunting and gathering,' she says.

So much for the statement that livestock guzzles too much water in comparison to plant food – the opposite is actually true when herd animals are sustained on natural vegetation alone and without cultivated feed!

# Cooling the Climate

*When properly managed, rangeland agriculture is fully sustainable, having gone on long before the discovery of fossil fuels, and it will, without doubt, go on long after the depletion of fossil fuels.*

R.K. Heitschmidt et al.[1]

The Food and Agriculture Organization of the United Nations (FAO) is the global authority on all aspects of food production. As behooves an institution of such august status, its reports are thoroughly researched and extensively referenced tomes which cover their subjects in dry detail or provide methodical hands-on instructions. They hardly qualify as a riveting read and are certainly not the kind of stuff that thrills your average urbanite or causes a ripple in the media.

But when the FAO published a report entitled 'Livestock's Long Shadow' in 2006, it was a bombshell. Its reverberations are still felt today – and will probably still be felt for a long time to come. The volume, compiled by a team around the FAO's chief livestock economist, Henning Steinfeld, made the startling statement that the livestock sector emitted more greenhouse gases than the whole transport sector combined, amounting to 18 per cent of human-caused climate gas emissions.[2] It did not stop there. The report also enumerated and detailed many other negative effects of the world's rapidly expanding livestock population on the environment, including on soils (erosion, desertification), air and water (heavy pollution), and oceans (dead zones).

In promoting the study, Steinfeld stated that '...livestock are one of the most significant contributors to today's most serious environmental problems. Urgent action is required to remedy the situation.'[3] The report warned that 'The environmental costs per unit of livestock production must be cut by one half, just to avoid the level of damage worsening beyond its present level.' It also elaborated: 'When emissions from land use and land use change are included, the livestock sector accounts for 9 percent of $CO_2$ deriving from human-related activities, but produces a much larger share of even more harmful greenhouse gases. It generates 65 percent of human-related nitrous oxide, which has 296 times the Global Warming Potential (GWP) of $CO_2$. Most of this comes from manure. And it accounts for respectively 37 percent of all human-induced methane (23 times as warming as $CO_2$), which is largely produced by the digestive system of ruminants, and 64 percent of ammonia, which contributes significantly to acid rain.' Finally, it added that 'Livestock now use 30 percent of the earth's entire land surface, mostly permanent pasture but also including 33 percent of the global arable land used to producing [sic] feed for livestock. As forests are cleared to create new pastures, it is a major driver of deforestation, especially in Latin America where, for example, some 70 percent of former forests in the Amazon have been turned over to grazing.'

The report was certainly correct in pointing out that the livestock sector is currently a major cause of environmental destruction. But it erred dreadfully in not distinguishing between the environmental impacts of the different systems of livestock keeping. Instead, by putting the whole range of production systems from pastoralist to industrial in the same bag, it gave livestock an overall bad rap. Moreover, 'Livestock's Long Shadow' heaped special scorn onto ruminants for the fact that they emit methane ($CH_4$). The so-called 'mono-gastrics' (poultry and pigs) were let off more lightly because they do not generate methane. One of the recommendations of the study was to switch to them for the sake of the climate.

The news was eagerly lapped up by the media – as well as by car manufacturers who now cheerfully advertised their wares as being better for the environment than a cow. The vegan movement has gloried in the data ever since. Widely watched films have spread the message with glee, cherry-picking data and not adjusting their messages when data were subsequently revised by the FAO, which reduced the percentage of livestock-caused Greenhouse Gas (GHG) emissions from 18 per cent to 14.5 per cent.

One of the surely unintended consequences of 'Livestock's Long Shadow' was that international aid agencies all but stopped supporting livestock-related projects because they feared they would be abetting global warming. This was a tragedy in itself since livestock is the most important asset not only of pastoralists, but also of hundreds of millions of poor smallholders in developing countries. Projects that improve livestock health are among the most impactful approaches to poverty alleviation in those places. But by means of this official UN report, livestock had practically become a dirty word.

Something else was set in motion: a search by battalions of scientists for measuring and minimising the emissions of greenhouse gases from livestock, with a special emphasis on methane. Addressing the urgent issue of livestock sustainability was practically reduced to its climate dimension, while other aspects of farm animal production, such as water and air pollution, biodiversity loss and the spread of antibiotic resistance, were given rather short shrift. Funding was poured into conducting 'life-cycle assessments' to compare the emissions of different livestock production systems per unit of product. When such a yardstick was applied, pastoralist systems came out the worst.

In order to address the problems identified in 'Livestock's Long Shadow', the livestock science community zeroed in on improving the 'efficiency gap' of livestock in terms of natural resource use. Natural resource use efficiency is defined as the amount of feed or nutrients consumed by an animal versus the amount of product – that is, meat or milk – it produces. The proposed solution was that we should 'make use of improved genetics, balanced feed and nutrition, and good animal health and husbandry which will help to reduce methane emissions per unit of output'.[4] 'Closing the efficiency gap' by making the livestock systems in developing countries as efficient as those in the US and Europe was the motto. Mainstream livestock scientists recommended that the best ways to mitigate GHG emissions was by intensifying production, basically applying the energy-intensive modus of production to the rest of the world.[5]

This approach comes with a set of fundamental problems:

Firstly, focusing solely on the amount of greenhouse gases generated per kilogram of product is reductive, as it ignores a livestock system's overall impact. It disregards the fact that herding systems also shift carbon from above ground to below ground, and thereby favours intensive livestock production that uses antibiotics as growth promoters and hormones to produce more meat and milk. The negative effects of this kind of production with respect to animal welfare, biodiversity loss, antibiotic resistance, and so on are externalised, that is, they are not considered.

Secondly, switching to high-input systems increases the use of fossil fuels and emissions of $CO_2$, as feed needs to be especially grown. High-yielding breeds need concentrate feed that has been grown using fossil fuels. They also need to be sheltered from the environment by installing air conditioning and ventilation.

Thirdly, recommending a switch to high-performing animals and 'improved genetics', and ignoring our pool of drought-adapted livestock breeds which require no external inputs, is dangerous in a climate-change scenario, as it undermines humanity's ability to adapt.

Fourthly, the feed requirements of 'improved genetics' compete with those of people and, given warnings by plant breeders that crop yields may decrease, are a burden that our planet may not be able to cope with. In an increasingly resource-constrained world, we can no longer afford to expend energy on growing feed for livestock on arable land.[6]

Finally, the nutrient density and quality of livestock products is not the same in intensive and extensive systems, so the amount of product is an insufficient proxy for its nutritional value.

While the FAO and mainstream livestock scientists recommended a switch to poultry and pigs in addition to livestock intensification as treatment for the problem, soil scientists make diagonally opposite recommendations: we urgently need to include ruminants in all agricultural cycles. They draw attention to the fact that grazing – whether by wild herbivores or domestic livestock – can actually mitigate or lessen the rate of climate change by improving the soil, increasing grass cover and stimulating root growth. They point out that grasslands are more potent 'carbon sinks' than forests. One of the reasons for that is their huge extent, amounting to about 5 billion hectares (12 billion acres), which is equivalent to about 70 per cent of agricultural land, as we have seen. And these grasslands can only be sustained if they continue to be grazed. In the absence of wild herbivores, livestock is the only choice.[7]

Professor Rattan Lal is one of the world's most respected soil scientists. He was a member of the Intergovernmental Panel on Climate Change (IPCC), which won the Nobel Prize in 2007, and has received numerous awards for his work in identifying agricultural practices that protect soils, prevent anthropogenic climate change and feed people. He founded the Carbon Management and Sequestration Centre at Ohio State University. Quoting Sir Albert Howard, a founder of the organic agriculture movement, one of his epithets is that 'the health of soil, plants, animals, people and ecosystems is one and indivisible'.[8]

A paper co-authored by Professor Lal with a team of well-known soil scientists suggests that ruminants have the potential to reduce global warming by enhancing the sequestration (absorption) of carbon in soils. The central argument is that cropping and soil erosion due to tilling cause 13.7 per cent of anthropogenic GHG emissions. By comparison, domestic ruminants emit 11.6 per cent of emissions but – if kept in grazing systems where they keep the sward and the grass roots system intact – they actually sequester more carbon than they emit. The researchers recommended including ruminant grazing in all agricultural management plans and cycles as a way of restoring nutrients to the soil.

Since 'Livestock's Long Shadow', the FAO, too, has realised the potential of grasslands for carbon sequestration and has engaged with its member states on this issue.[9] However, the amount of carbon that can be sequestered remains a controversial subject on which experts simply do not agree and which is very much debated. It appears to vary from place to place and is also very much a question of skillful and controlled grazing.[10]

In parallel to the discussion about the impact of domesticated animals on carbon sequestration, the role of large herbivores such as elephants and wild boar in the carbon cycle is also being examined. Through their manure and urine, they cycle the carbon contained in plants from above the ground into the soil where the biomass is decomposed by microbes and earthworms. Carbon is much more stable below the ground than above, where it is prone to fires if not grazed.[11] Simply put, plants produce energy via photosynthesis. Herbivores are necessary to dissipate that energy, and if they do not do it, then fires will occur. Herbivores are also necessary to cycle plant carbon below ground, where the plant material decomposes to become available again for other plants.

In the meantime, the whole edifice built around achieving livestock sustainability by reducing methane from livestock production appears to be crumbling, as was elaborated in a recent report collated for the Climate Conference in Glasgow.[12] For one, questions have arisen about the differential impact of the three main types of greenhouse gases. $CO_2$ is emitted whenever we burn fossil fuels in agriculture, fertiliser production, transportation and most human activities. It's a long-lived pollutant that builds up over centuries and lasts for thousands of years. The $CO_2$ that was released by burning coal in the eighteenth century is still having an effect today. It is cumulative, and even if we stopped producing it today, the effects from what we did two hundred years ago would still be there.

Methane is produced naturally by decaying plants in anaerobic conditions and when it is emitted by domestic ruminants, it is entirely biogenic – it is derived from the $CO_2$ that plants absorb from the air to build their structures, that ruminants then ingest and that the bacteria in ruminant stomachs finally break down into sugars that then build ruminant bodies – a very important process that makes it possible to produce protein in uncultivable land. So, it is part of the natural carbon cycle of which all ruminant animals – whether wild or domestic – are a component. Methane's potency is much higher than that of carbon dioxide. But it is short-lived, lasting only about 10–12 years, so if the ruminant population does not increase, methane emissions also remain stable.

The most potent greenhouse gas of all is nitrous oxide which has three hundred times the warming potential of $CO_2$ and lasts for over a century in the atmosphere. It also destroys ozone in the stratosphere. While it occurs naturally, isotope studies have confirmed that its increase is mainly a consequence of Green Revolution agriculture, which is based on high-yielding crop varieties in combination with chemical fertilisers. Scientists recommend minimising the use of chemical fertilisers for this reason, but also acknowledge that this impinges on food security.[13]

The effect of the three gases is conventionally expressed in a metric called Global Warming Potentials (GWP) over a hundred years, but this metric does not take into account that methane does not last for 100 years. In order to address this issue, scientists from Oxford and New Zealand have developed a new method that considers the different lifetime effects of greenhouse gas emissions.[14] Farmers' organisations are currently lobbying the IPCC to adopt this metric in order to reflect the climate impact of grazing systems more accurately.

Secondly, the vast majority of Life Cycle Analyses (LCA) have been undertaken in the Global North and focused on intensive production.[15] Results have then been inferred globally, assuming the same parameters apply to extensive systems when the few available studies actually indicate that this is not the case.

Another argument has been brought forward by Pablo Manzano, a Spanish ecologist. Even if we removed pastoralist livestock from rangelands, this would not reduce the emissions of methane, as wild ungulates and termites would be expected to move in and then emit an equal or even larger amount of this greenhouse gas.[16] For North America it has been estimated that the bison, elk and deer that roamed the land before European colonisation emitted methane to the tune of 86 per cent of the modern cattle population.[17]

In sum, the only real solution to reducing greenhouse gases and stopping the rate of climate change as a result of agriculture is to minimise the use of fossil fuels as well as reduce the exposure of soil to the elements by avoiding ploughing. Large-scale crop cultivation depends on the use of heavy machinery as well as chemical fertilisers. By contrast, pastoralists are powered by the sun and never leave the soil bare, keeping it well covered with plant material.

---

While the question of which kinds of livestock systems contribute what to climate change or climate stabilisation is sure to keep scientists on their toes for quite some time, here is some practical information on how herded animals can assist us in coping with events that climate change is set to make more frequent and more destructive: floods and fires.

In mid-July 2021, torrential rains – 148 litres per square metre (33 gallons per 11 square feet) within 20 hours – caused a catastrophic flood in the Ahr River valley in the Eifel mountains of Germany, killing more than 170 people and leading to the destruction of bridges, buildings, roads, even whole towns, which will take decades to rebuild. According to Günther Czerkus, the president of the German professional shepherds, a major factor leading to the catastrophe was the fact that the soil in most of the affected area had lost its capacity to absorb water because of compaction and lack of humus. He notes that the valleys that are too steep for cropping experienced hardly any damage, whereas the valleys without intact grassland experienced the worst damage. In his opinion, the difference is due to grazed areas having an intact soil life and therefore a high water-absorption capacity, while the cropped soils do not have this, in addition to being very compacted due to the use of heavy machinery.[18]

While Günther's observations will need scientific validation before being accepted and acted upon by policy makers, the role of grazing animals, especially sheep, in maintaining dykes are undisputed. Dykes are a major tool in preventing the flooding of lowlands from the sea and rivers. Grazing animals maintain their functionality by keeping the grass short and compressing the soil with their hooves. In the absence of sheep, dykes need to be mown but this has negative effects on insect life, besides requiring fossil fuels. Grazing is also superior in creating the dense grass cover that holds soil in place. The coast protection service in Germany is heavily reliant on shepherding for its work and there are dozens of professional dyke shepherds who take care of almost all of Germany's coastline and the embankment of the Elbe River, combining environmental services with the production of high-quality lamb.

With respect to fire prevention, the role of grazing animals in removing plant mass is unquestioned. Herding provides a very much sought-after means of preventing wildfires. Grazing goats prevent the greenhouse gas emissions that fires cause and does not leave the soil bare. The fire department of California has used goats for that purpose for about a decade and a number of businesses deploying both sheep and goats have sprung up. Demand for grazing goats is growing like wildfire, according to the Sierra Club, the most respected nature conservation organization in the US. In October 2019, a herd of goats saved the Ronald Reagan Presidential Library from going up in flames. One of the pioneers in the field is Lani Malmberg who started the Goatpelli Foundation, a non-profit organisation that teaches people how to deploy goats for fire prevention. Born on a cattle ranch in Nebraska, Lani studied environmental restoration and also added a degree in Weed Science, only because stipends were available for this subject. This qualified her to get a shepherding job on a ranch that sought to control weeds without herbicides. This experience led her to set up her own business in 1996 by buying a hundred goats. Now she has more than a thousand of them and is deploying her goat families in 17 western states to take care of land, mostly for fire control. 'The fear of fire is now so strong, that I currently have 300 open requests for grazing my goats,' Lani told me. She manages the goats, who now represent the 27th generation of the ones she initially purchased, with the help of a family of Border Collies and sometimes portable electrical fencing. Lani emphasises the advantages of goats over dumping water by helicopter: goats do not just remove the brush, they also build soil organic matter which improves water-holding capacity and thereby prevents wildfires, while just dumping water does nothing for the soil.[19]

This approach works not only on ranches in the Western USA, but also in cities further east. In Ohio, there is an initiative involving urban shepherds that promotes using sheep to replace lawnmowers. Instead of having lawns maintained by mechanical mowers, which costs about US$1,900 per year an acre, it is much more sensible and ecologically sustainable to have sheep perform the job. An acre (4,000 sq m) produces about 272kg (600lb) of grass which can be converted into 73kg (160lb) of meat and several pounds of wool, in addition to 272kg (600lb) of manure for plant growth. The pollution caused by a lawnmower in the form of $CO_2$ and $NO_2$ emissions can be avoided, unwanted invasive species controlled, and native grasses restored. All by cute and cuddly sheep!

There is more positive news from the other end of the world, where reindeer herding engineers ecosystems and protects us against global warming.[20] The Arctic is the region most affected by global warming. This is reflected in a process called Arctic greening which means that the original lichens are replaced by shrubs and eventually even trees. In a vicious feedback circle this further increases global warming, as shrubs and trees absorb more solar energy. Studies in Russia's Yamal Peninsula, which is inhabited by the Nenet reindeer herders, suggest that grazing by reindeer can stop shrubification and prevent the tundra from turning into a wooded taiga. As the authors of the study explain:

> Our results thus point towards increases in large herbivore pressure having compensated for the warming of the Peninsula, halting the shrubification of the area. This suggests that strategic semi-domesticated reindeer husbandry, which is a common practice across the Eurasian Arctic, could represent an efficient environmental management strategy for maintaining open tundra landscapes in the face of rapid climate change.[21]

Calls to end animal agriculture to reduce GHGs persist. This reductionist thinking ignores how crucial animals are in the streams of energy and nutrients that are the basis of life and all biological processes on Earth. We need animals to dissipate the energy that plants have generated from the Sun. In the earth cycle, animals ingest this energy to grow, move and reproduce. They return waste products to the soil and the cycle starts anew. In the absence of animals, the energy accumulated by plants lends itself to fires and the nutrients then disintegrate and evaporate. Herding people support their animals to turn 'waste' into nutritious food and to recycle unused nutrients into the soil, more or less mimicking the role of wild herbivores. They demonstrate how animals and people can jointly set things right for the planet.

# Feeding the World

*Where agriculture in arid lands is impossible nomadism is not only the best way, but it is the only way of using the land for man's benefit.*

HILDE GAUTHIER-PILTERS and ANNE DAGG[1]

How will we feed the almost 10 billion people that are projected to crowd our Earth by 2050? Especially in times when the climate is getting hotter and weather is becoming more unpredictable? What is the role of pastoralists in providing adequate food for the burgeoning human population?

Manufacturers of artificial meat and dairy, vegans and quite a few scientists insist we should stop eating meat and convert to a plant-based diet. The arguments go like this: '…in the US an acre of land is needed to produce 250 pounds of beef, in comparison, the same amount of land has the potential to produce 50,000 pounds of tomatoes or 53,000 pounds of potatoes', and continue: 'By some estimates, we could feed 1.4 billion additional people simply by giving up beef, pork, and poultry in the United States. Think of what we could do if the entire world gave up all animal products!'[2]

Another favourite statistic regularly trotted out, including by bona-fide scientists, is that raising livestock requires an enormous amount of space but provides very little in return. 'Meat, aquaculture, eggs, and dairy use 83% of the world's farmland and contribute 56 to 58% of food's different emissions, despite providing only 37% of our protein and 18% of our calories,' according to the aforementioned article mentioned in the earlier chapter, *A Tapestry of Cultures*, published in the journal *Science* and copiously quoted by mainstream media.[3]

However, such statements are utterly misleading. As I discussed earlier, the Food and Agriculture Organization of the United Nations (FAO) classifies a little more than a third, or 37 per cent, of the global land mass as 'agricultural land'. But 'agricultural land' does not mean that crops can be grown there. Only about a third of agricultural land, or 11 per cent of the global land mass, is considered arable cropland. In the remaining two-thirds of agricultural land (equivalent to 26 per cent of the global land mass), it's too hot or too cold, the terrain is too steep, the soil too stony or too salty, the rainfall too little or too much, or the growing season too short. These areas are referred to as 'permanent pastures' in the FAO classification system. If people want to produce food there, they can only do this by means of livestock. Just imagine: more than two-thirds of agricultural land can only be used for food production by raising animals.

As we have seen, it is not the fault of livestock that it uses so much land. It is simply because our farm animals have a much wider ecological range than cultivated plants, meaning they can move around and produce food in huge areas of the globe where no crops grow, from the freezing Arctic to the scorching Gobi Desert and from Himalayan heights to the Danakil Depression below sea level. On no continent is this more important than Africa where its 268 million pastoralists occupy 43 per cent of the land mass and contribute between 10 and 44 per cent of the GDP of individual countries.[4]

The statement that livestock provide only 37 per cent of our protein and 18 per cent of calories is based on calculations by the FAO. It is probably accurate, but the insertion of 'only' is misleading. Clearly, it is not the function of animal-sourced food to provide us with calories; these are much more easily obtained from starchy, carbohydrate-rich food such as cereals. Globally, there is no shortage of carbohydrates and obesity as over-consumption of calories is much more common than a lack of calories. Hunger and malnutrition, on the other hand, are caused by a lack of protein and of micro-nutrients rather than of carbohydrates. Animal-sourced food happens to be much more protein-dense than any plant food, in addition to coming with a more complete range of amino acids. Even relatively small amounts of animal protein enormously benefit young children living in food-insecure areas, reducing stunted growth and resulting in higher achievement in schools. Dietary deficiencies in vitamins A and $B_{12}$, as well as in iron, iodine, zinc and folic acid – which are common in both pregnant women and children, especially in poorer regions – can be best alleviated by means of meat and dairy. Animal-sourced food not only has much higher quantities of these micro-nutrients, but also much better bioavailability compared to plant foods, meaning the nutrients can be more easily absorbed from the gut and metabolised.[5]

And here is the beautiful thing about pastoralism: it is by far the most efficient way of producing protein, since ruminants such as cattle, sheep, goats, yaks, buffaloes and camelids have the ingenious ability to synthesise protein from very fibrous vegetation and from crop residues that are inedible for humans. They do this with the help of the micro-fauna in their fore stomachs which digest cellulose-rich forage and resynthesise it into amino acids. Pigs and poultry, who are mono-gastrics (meaning they have only one stomach), have a different metabolism and, as omnivores, can convert a variety of food waste into high-value protein.

While pastoralists deploy livestock to transform 'waste' into nutrient-dense food without any use of fossil fuels, in 'modern', 'scientific' livestock production, animals are fed with soy, maize, wheat and other nutrient-dense cereals grown especially for them in huge monocultures that often extend to the horizon. Pigs and poultry receive carefully calculated rations composed of grains. In dairy operations and feedlots, cattle are stuffed with barley, maize, sorghum, wheat and oats that could be directly consumed by people. This is not good for the cattle, whose metabolism is not adapted to this kind of feed and who therefore require drugs to keep them alive. It is also absolute nonsense from a food security perspective, as in this way more protein is fed to animals than produced by them. It is an utterly wasteful approach.

This is testified to by Henning Steinfeld, chief of the livestock policy section of the FAO and the initiator of a global debate on the environmental impacts of livestock as well as of the Global Agenda for Sustainable Livestock (GASL), a platform through which governments, researchers, industry, NGOs and livestock keepers discuss how to move towards more sustainable livestock production. At a meeting of livestock donors held at the offices of the Bill and Melinda Gates Foundation in Seattle in 2014, he included a slide in his presentations that compared the protein efficiency of livestock systems in different countries. Protein efficiency measures the amount of human-edible protein consumed by livestock versus that produced by livestock. Steinfeld's table clearly shows that the countries with supposedly 'backward' pastoralist systems are absolutely superior in this respect. Kenyan and Ethiopian livestock, which is predominantly raised by pastoralists, come out on top, producing up to twenty times more protein than is fed to them. By contrast, the livestock of the US and Saudi Arabia, where livestock is raised mostly in feedlots, consumes more than twice as much protein than it produces. In essence, industrialised livestock systems destroy protein since the animals are fed on grains that could also be eaten directly by people. Recent research from Australia came up with even more astonishing figures, concluding that cattle raised exclusively on pasture produced 1,597 times more protein than they were fed, but if they spent their last months in feedlots where they were fed with grain, this dropped to less than two.[6]

If we look at the future needs of feeding the world in 2050, it is unlikely that we will be able to afford to grow special fodder crops for 'high-yielding' animals, as this entails a huge consumption of energy and loss of protein. Instead, we must make the most of the ability of ruminants to upcycle naturally growing vegetation, which cannot be consumed by people, into milk and meat. In a resource-constrained world, livestock systems that can make do with 'waste' growing in far-flung and remote places are an invaluable asset.

In this context, a large-scale applied research project in the US to combine both objectives by re-establishing the original grasslands in a grain-growing belt in the US is of major interest. North America's prairie is a unique grassland ecosystem that co-evolved over many millennia with several large ruminants, especially buffaloes but also deer, elk and antelopes. Composed of a large number of both short and long grasses as well as flowers, it was an ecosystem remarkably resilient against frequent droughts, due to the extensive underground root system. It formed a rich hunting ground for Native Americans who used fire to manage this ecosystem. White settlers not only hunted the buffaloes to extinction, but gradually converted the prairie to cropland, cultivating it with maize and other food plants, so that currently only 0.1 per cent of the native vegetation remains. Huge feedlots or Concentrated Animal Feed Operations (CAFOs) have been set up in which cattle are fattened with grain such as maize. This system yields large amounts of meat, but causes huge collateral damage in the form of soil erosion, nutrient loss, water pollution, greenhouse gases and the extinction of biodiversity. Nor is it profitable for farmers, many of whom have been driven out of business, provoking the question of why cattle cannot directly feed on grass, and why this whole detour via cultivation and fattening with enormous inputs in terms of fossil fuels and fertilisers is necessary. With support from the United States Department of Agriculture (USDA) for a long-term project, a team from the University of Wisconsin is now looking into restoring the ecosystem and transforming the grain-based livestock production to one that is perennial grass based. The hope is to set up a sustainable system that is also profitable for farmers in an extended process which involves stimulating demand for grass-fed beef. The project is called Grassland 2.0.[7]

Scientists refer to farm animals that can utilise low-grade, fibre-rich forage as 'low-cost livestock', meaning inexpensive to feed and keep. A team of researchers from Wageningen University in The Netherlands has demonstrated the importance of low-cost livestock for food security by utilising and transforming the by-products of crop cultivation into food. If we grow wheat, we do not just end up with grains, but also with straw. Low-cost livestock can metabolise this straw and convert it into meat and milk. In the absence of livestock, the straw would just go to waste. The researchers have calculated that if we switched from high-yielding to low-cost livestock in Europe, we could still produce enough protein to cover the needs of Europe's inhabitants and become independent from feed imports.[8] But because we currently favour high-yielding breeds, we depend on massive imports of soya beans from South America – 70 per cent of the protein needs of European livestock is covered by imports.

The collateral damage of the current system is significant. On the one hand, it leads to massive overproduction of meat and milk that is exported with subsidies and at low prices to other countries, where it undermines the livelihoods of the local livestock keepers. On the other hand, soya bean cultivation in South America is one of the driving factors in the deforestation of the Amazon in Brazil and the ploughing up of the pampas in Argentina. In addition, the manure of the confined animals fed with that protein accumulates in huge quantities instead of being dropped onto the soil in therapeutic doses by herd animals. Manure from industrial systems turns into a toxic mass that pollutes groundwater with nitrogen, besides releasing greenhouse gases into the atmosphere.

Unfortunately, many farmers have been thoroughly brainwashed by the array of industries (fertiliser, feed, agricultural machinery, genetics) that benefit from this wasteful, unsustainable system into believing it is their duty to 'feed the world' and are unwilling to reconsider their ways of producing food. They present a powerful resistance against any efforts to put some sense into the debate about what constitutes livestock sustainability.

The team from Wageningen University also analysed what would happen if industrial livestock farming and feed cultivation stopped, and we instead limited ourselves to feeding livestock on either crop by-products and food waste or by grazing them on non-arable land. Their conclusion is that if we combined this approach with reducing the consumption of animal-sourced food in Western countries, production would still suffice to increase protein availability in Africa and Asia. They also surmised that sustaining the human population on plant food alone would actually require MORE land because, without livestock, crop by-products would not be utilised for food and so go to waste.[9] If we wanted to eliminate animal-sourced food from our diets, we would need to expand the area under crop cultivation to replace the nutrients currently provided by meat and milk. And metabolising farm waste is hugely important for preventing further greenhouse gas emissions, as its incineration or decay in a landfill would generate additional greenhouse gases. If livestock consume crop by-products, such as citrus pulp, almond hulls and distillers' grains, they contribute to food production while also reducing climate gases.[10]

Despite this evidence, the high-input-high-output systems of livestock production are the model that most developing countries aspire to. But an appreciation of pastoralism is gradually expanding, and Africa is taking the lead. The African Union has come up with a pastoralist policy framework stating that 'over 95% of the inputs for traditionally reared, extensively grazed ruminants come from the sun, and soil, and cost the producer very little'. It recognises that the 20 million people dependent on pastoralism would have to find alternative sources of livelihoods if this way of life ceased.[11] The FAO, too, is taking an increasing interest in pastoralism.

---

Returning to my opening question of whether pastoralism can feed the world, the answer is yes, it can. It would certainly not produce as much meat and milk as the current high-input industrial systems do. Meat and milk would become rarer and cost more, but we would consume it in tune with 'planetary boundaries' or within the limits imposed by the Earth. We would save enormous amounts of fertiliser, conserve biodiversity, prevent air and groundwater pollution, significantly curb the use of antibiotics and prevent billions of animals living miserable lives if we made the transition to herding and other forms of extensive livestock keeping, with an emphasis on maintaining herding in remote areas. Eliminating industrial systems of animal production while supporting herding and keeping livestock on locally available resources is ecologically more sustainable and a realistic option we should strive for.

Pastoralism is an essential component of providing healthy and sustainable diets to the growing human population. It synthesises protein without depending on fossil fuels and without requiring monocultures. Moreover, it does not compete with crop cultivation but instead adds value to it. Quite apart from that, herding is the one and only option for food production in the marginal areas of the world.

It is a personal choice to be vegan. But it is ecologically feasible only in the relatively small temperate parts of the world where crops can be grown. For both cultural and ecological reasons, it is an option that cannot be imposed on the rest of the world. In rural Mongolia, for instance, no cereals and vegetables, let alone fruit, can be grown, leaving animal-sourced food as the only alternative – and the cuisine that people have depended on since millennia.

For those of us who have the luxury of choice, limiting our consumption of animal-sourced food to what can be produced in grazing and pastoralist systems makes most sense. The concept of largely 'plant-based diets' is sound in principle, but unfortunately it has already been co-opted by the multinational companies that control the global food sector and earn from highly processed food. The solution is to shun industrial food for what can be locally produced in tune with available resources, whether it is from plants or from animals.

# Balancing Nutrition

*A healthy body knows what to do regarding foods and diets, given appropriate choices and social models.*

Fred Provenza, Michel Meuret and Pablo Gregorini[1]

In February, our itinerant camel herd feeds on the camel thistle or *unt-kantalo*, officially known as the Indian Globe Thistle, a waist-high, very prickly plant with spiky, baby-blue, spherical flowers that are reminiscent of the coronavirus – or micro versions of our blue planet. Botanically, it is referred to as *Echinops echinatus* and it grows during the dry season on farmers' fields in the furrows left from the plough. It proliferates all by itself, covering the ground with a dense layer of thorns that makes it difficult to walk through the fields. The farmers hate *kantalo* with a passion; for them it is a terrible weed that makes it difficult for them to seed the next crop. But camels love *kantalo*, so the Raika take their herds to de-weed the *kantalo* fields, which pleases both the farmer and the camels.

Visiting the herd at this time of year is a priceless experience – February is the month when most camel babies are born. While their mothers are browsing, the new arrivals are tottering around, trying to coordinate their elongated legs that make up about 80 per cent of their body mass. Gradually they become more sure-footed and within a few weeks, they gallop around and form friendships with their batch mates. In their first few days, they may need a herder's helping hand to stand up and suckle milk. At this stage, the females usually have far more milk than the calves can ingest, so it's a good time to taste camel milk, which is traditionally drunk from a folded leaf of the *aak*, a small tree that grows widely on low-nutrient soil and is avoided by livestock because it has a very bitter, latex-like sap that is toxic. However, the unblemished leaves lend themselves to being folded into cups and the Raika usually carry some of these in their turbans to enjoy a drink on the go.

I am not generally a milk drinker, but this milk is something else. The camel thistle diet renders it extraordinarily sweet and, when it is freshly milked and runs down your throat at body temperature, it feels like manna, so smooth and satisfying that camel herders can live on it for weeks, if need be.

We can also assume that camel milk is extraordinarily healthy. Ancient Ayurvedic texts describe the Indian Globe Thistle in detail, enumerating its healing properties, many of which have been confirmed by modern medical research. The plant is held to be antifungal, analgesic, protective of the liver, antioxidant, anti-inflammatory, wound-healing, fever reducing, antibacterial and especially good for your sex life.

---

Camel milk is one of the most intriguing pastoralist-produced foods and is ascribed a variety of therapeutic qualities. Its consumption was even held responsible for the once diabetes-free status of the Raika community.[2] Further research has shown that when the Raika stop camel herding, they are especially prone to diabetes.[3] To me it is questionable whether camel milk was the only factor behind the absence of diabetes among the Raika camel herders – the change in lifestyle from walking 20–30km (12–18 miles) per day to a sedentary existence will also have played a role. Nevertheless, camel milk has been shown to make a difference to city-based diabetes patients, enabling them to cut back their insulin needs by 30 per cent.

Although regarded as exotic by consumers in countries without a tradition of camel herding, it is rapidly gaining a devoted customer base in the US, Australia, the Far East and even Europe. Camel milk is compatible with the huge number of lactose-intolerant people who cannot drink regular cow's milk. It can also have almost-miraculous impacts on autistic children.[4] Referring to recommendations made by Prophet Mohammed, Arab and Muslim cultures believe camel milk to be helpful in combating cancer, especially in combination with camel urine. So, health reasons are a major factor in its growing popularity.

A number of large camel dairies in Dubai and Abu Dhabi are trying to cash in on this and have set up huge farms with thousands of camels. Although a conscious and admirable effort is made to create humane systems in which babies spend time with their mothers and there are large exercise tracks for the camels, the diet of the camels is industrial – as the farms are located in the middle of the desert, feed for the camels must be imported from places like Canada and Australia.

Here in Rajasthan, we are trying – struggling – to implement a different vision. A system in which camels are purposefully kept in the traditional nomadic system, so they can continue to feed on the 36 Ayurvedic plants (more on this later) and in which the flavour of the milk changes with the seasons. And so the camels can continue to perform their ecological functions of fertilising cropland, disseminating acacia seeds and 'de-weeding' fields while savouring *kantalo*.

The Raika camel milk has been included in the 'Ark of Taste', a collection of heritage foods deemed to be endangered and that the International Slow Food movement has set up to highlight foods linked to specific eco-regions that are produced traditionally and have a distinctive taste. The milk makes utterly delicious cheese which is popular among Rajasthan's royal families and heritage hotels. It is a quadruple win-win situation for camels, Raikas and the environment. But it would all fall apart if nobody wanted to buy the milk.

---

Grass-fed beef, organic milk and eggs from free-ranging chickens are all the rage in the US and Europe among those who are well to do and not vegetarian or vegan. Nutritionists confirm that these products are not only more animal-welfare friendly but also much healthier than those from confined animals: they are leaner and their composition is more conducive to our physical well-being. They also have higher proportions of omega 3 rather than omega 6 fatty acids.

Yet, 'poor' rural people in India are equally aware of the importance of healthy food and often prefer a more expensive local product over a cheaper, anonymous one. The redoubtable Dailibai Raika may be officially 'illiterate', but this does not stop her from understanding the connections between healthy soils and the nutritional quality of food. She aspires to buy only those vegetables that have been grown with *deshi khat*, meaning animal manure. Dailibai, and many other Raika and people who have never been to school, frequently lecture me about such matters. They point to the epidemic of high blood pressure, cancer and diabetes that now affects many Indians and link this to the use of chemical fertiliser and modern, fast-growing crops. Their instincts about what is healthy food are true.

Pastoralists and other local livestock keepers are very much aware of the nutritional value of their produce and often prefer to consume it themselves, rather than sell it for a high price to others. This is another lesson learned from working with herders in Jaisalmer. This desert district is considered the poorhouse of Rajasthan and regarded as one of its most backward areas (and in terms of gender equity that is certainly true). But its people go to great lengths to obtain *deshi* (local) *ghee* (butterfat) from local cows, rather than purchase a packaged variety in a shop for half the price. When you ask them why, they will reply something like: 'It's much healthier because of the grasses that the local cows eat.' The same applies to meat. The local meat shops where goats are slaughtered on demand sell Jaisalmeri goat meat at a significantly higher price than that from non-local goats. Yet, on the infrequent occasions the villagers eat meat, they will go for the pricier local option.

Probing around to find out why exactly the local produce was thought to be superior to that from elsewhere, I came across the concept of the 'thirty-six plants'. This was a belief that both goats and camels forage on 36 plants that have certain medicinal qualities, rendering the animals and their products extremely healthy. The people in the Thar Desert had difficulty fathoming that I was not familiar with this. 'How can you not know about such basic facts of life,' their looks implied, when I was questioning them to find out more.

Seeking to make up for our ignorance, we tried to identify the 36 forage plants from our learned informants…and found out that the plants they named differed to varying degrees. While our respondents agreed on about 15 of them, there was no consensus on the remaining ones. It became clear that camels – and probably goats – actually browse on far more than 36 different shrubs and trees. So 'thirty-six' is just a turn of phrase that indicates wide diversity. Interestingly, almost all the plants identified are used in traditional medicine and are referred to in the Ayurveda, India's ancient body of medicinal knowledge. Many of them have also been analysed scientifically for the presence of substances with medical or therapeutic values. One example is jujube (*Ziziphus nummularia*), an extremely drought-resistant and very thorny shrub that belongs to the Rhamnaceae family. Its various parts – fruits, leaves, roots – are mentioned in the Ayurveda as being therapeutic for a range of afflictions. Modern scientific research has shown its extracts have antibacterial and cancer-inhibiting properties on animal cells that have been cultured in a petri-dish.

As herded animals feed on very biodiverse forage, often composed of not only 36, but hundreds or even thousands of different plants, their diet is certainly entirely different to that of stall-fed farm animals who are given a small number of feed stuffs which often include genetically modified soya beans and corn.

Dr Stephan van Vliet, a nutritional physiologist currently at Utah State University, compared the composition of beef samples from cattle raised in a Midwest feedlot with those from mountain pastures in Idaho. With the help of mass-spectrometry, he conducted what is called a metabolomic analysis of the samples – that is, he looked for metabolites or biochemical substances that are produced by cells during metabolism. These can serve as an indicator for the physiological status of an organism. One category of metabolites are phytochemicals which are produced by plants in response to sunlight, moisture, soil nutrients, the surrounding plant community and grazing by animals.

The differences in composition were so dramatic that the two types of meat seemed like different substances. The phytochemical richness of grass-fed beef, from cattle grazed on plant-diverse pastures, was about three times higher than in the feedlot samples. The grass-fed beef contained more metabolites, which work as antioxidants. Antioxidants are associated with a number of health benefits, including a reduced risk of cancer and heart disease. The feedlot beef had a higher proportion of substances that are associated with heart disease.

Labels on packaged food provide us with information on the amount of primary nutrients, such as fat, carbohydrates and protein, as well as vitamins. But, as Dr van Vliet explained to me in an interview, the devil lies in the detail, and it is not only protein as such that is important, but the type of amino acids it is composed of. The nature of the phytochemicals in our food influences our body processes and can help protect us from diseases, such as heart problems and even cancer, or render us more susceptible to them. Chemically they include alkaloids, phenolic acids, flavonoids, glycosides, saponins, polysaccharides, stilbenes and tannins, and many of them have antidiabetic properties. There are at least a thousand different phytochemicals and so far only a few of them have been explored.

Phytochemicals are the secret sauce that keep humans and animals healthy, while also being an indicator of the health of the land on which food is produced. We are beginning to understand these connections between the environment, animals and people through the pioneering work of Fred Provenza, a professor at Utah State University who has spent his career looking at the feeding patterns of ruminants. In collaboration with his students, he found out that feeding behaviour is passed on from mothers to offspring, starting in utero, and that what animals eat in their youths determines their preferences later in life (as is also the case in humans).

In the 1990s, Fred paired up with Michel Meuret, an animal scientist in France, who studied the grazing circuits of French shepherds. In their monumental oeuvre *The Art and Science of Shepherding*, edited with other colleagues, they zero in on the role of the 'palate' in adequately meeting the nutritional needs of both animals and people. They posit that we are linked to our surrounding landscapes by our palates, or sense of taste, and the ability to distinguish and appreciate different flavours. This sensory capacity is developed while still in the uterus as well as early in life, as body cells send chemical signals to the brain in response to phytochemicals in the blood. The evolutionary purpose of the palate is to ensure that we get the nutrients we need.[5] Fred Provenza and others have shown that animals given the choice between a variety of forage plants choose diets that are suited to their individual needs. Studies among several animal species demonstrate they self-medicate and purposefully seek out plants that help them overcome diseases. For instance, goats with a high load of internal parasites eat plants that are rich in tannins and act as de-wormers.

Since modern diets have become very poor in phytochemicals, people, as they grow up, no longer develop the kind of 'knowledgeable' palates that can keep them healthy, but instead form palates that crave sugar-rich and fat foods. As modern agricultural practices emphasise weight, uniformity and transportability of produce and rely on significant amounts of chemical fertilisers, our cereals, vegetables and fruits have lost much of their nutrient density, so now our food must be artificially 'enriched' or fortified by adding vitamins, iron and other substances. The absence of satiating micro-nutrients in our diets makes people overeat and become obese.

If we reflect on how most livestock is raised – in confinement, not able to exercise, stuffed to the hilt with enormous rations of nutrient-rich feed and without opportunity for meaningful social interaction – we shouldn't only have a bad conscience about how these animals lived, but must also question how healthy their products are. In order to prevent animals from ailing and falling sick as well as to increase growth rates, antibiotics and hormones are a routine ingredient of livestock diets in many countries. Without them, these intensive farming systems would not function.

If we want animals to provide us with healthy food, they themselves need to be healthy and, very preferably, happy as well. For that they must be kept busy by the challenge of searching around for their own feed or favourite nibbles, instead of having a calculated ration placed in front of them, the same boring meal every day. As herd animals, they need to have the opportunity for a social life as well. Pastoralists provide such environments, which combine care with challenges. Although their animals have to cope with environmental stresses, fluctuating availability of feed and will eventually be eaten, theirs is a real life with the ups and downs we humans also experience.

And this is how herding connects our own health with that of the planet. It's a way of producing food that is natural and does not interfere with the planet's ecological functions. It produces nutrient-dense food that is rich in phytochemicals and keeps humans away from eliminating biodiversity and replacing it with monocultures. Obviously, we cannot go back to pre-agricultural times of living from hunting and gathering because our population size exceeds that capacity. But we can make optimal use of the planet's grasslands through skillfully managed grazing in a 'one-health' approach that not only keeps ecosystems healthy, but also animals and humans.

---

Food is not only about nutrition and health, but also about taste, of course. The most expensive ham in the world is produced in the *dehesas* of Spain and Portugal, the ancient landscapes created by livestock grazing in woodlands, especially oak forests, which are hubs of biodiversity – those that Jesús Garzón-Heydt is seeking to protect (see *Stewarding Biological Diversity*, page 133). In an ancient practice, small and slow-growing pigs of the Iberico breed are herded into these biodiversity-rich mosaics of meadows and ancient trees, where they feed predominantly on acorns as well as aromatic herbs. Their ham is a much sought-after gourmet speciality priced at up to 3,600 euros per leg. Ironically, several decades ago, this traditional, agri-silvi-pastoral system almost disappeared due to the mainstream promotion of fast-growing, stall-fed hybrid pigs that economists and efficiency believers are so fond of. Fortunately, the demand by culinary experts and gourmet cooks for the tasty meat of the Iberico pigs saved the breed, the habitat and a unique gastronomical experience. It also provided a test case for how everything is interlinked: the vegetation, the management system, the breed, the quality and the taste of the food.

It's a case study that cries out for replication. This has happened in Argentina where meat from goats kept in traditional transhumant systems in the mountains of Patagonia is high in demand. *Chivito criollo del norte neuquino* is an officially recognised 'denomination of origin', which means that only goat meat produced according to strict criteria in the traditional transhumant system can be branded as such. About 1,500 Crianceros on horseback accompany some 350,000 Criollo goats that move seasonally from lower-lying hot desert areas to higher and cooler places with more rainfall in Neuquén Province. The official recognition and support came about as a result of extensive research by scientists from Argentina's National Agricultural Research Institute (INTA) who not only documented this traditional system, but also sat down with the Crianceros in a truly participatory approach to work out and agree upon the production standards for the goat meat they produce. The denomination of origin for goat meat from Neuquén was officially approved in 2008 by a government resolution and President Cristina Kirchner herself visited the Crianceros to announce this.[6]

---

By selectively purchasing meat and dairy products that come from herding systems, we support rural livelihoods, biodiversity conservation and livestock welfare, besides doing something for our own health. Currently, there are no specific labels at the global level that can tell us whether a product is pastorally produced. Marking a product as pastorally produced would indicate even higher and more comprehensive standards than 'organic', which testifies only that livestock has been raised on organically grown feed and meets certain animal welfare requirements, but not that they are kept in a biodiverse system and as part of the landscape. We urgently need to develop such a label because it would give global visibility to herding systems!

# Death

*We need to learn to accept the death of animals,*
*as it is a part of the eternal cycle.*

RUTH HÄCKH[1]

Slaughter is not a subject for squeamish souls, and I can understand anybody who wants to skip this chapter. I feel the same. However, if we are serious about animal welfare we need to engage with the issue, like it or not. Avoiding it just multiplies animal suffering. If we hope the issue will fade away by loudly clamouring for the end of slaughter and pan-veganism, we are sticking our heads in the sand. And making the situation even worse for farm animals.

We can break down the subject into two parts: the necessity of slaughter and how we go about it. If we concede that a certain number of farm animals are essential for human nutrition and the health of the planet, we must accept the fact that culling is a biological necessity. As their guardians and caretakers, we have to take on the role of the apex predator in the system. In nature, predators play an ecological role, eliminating old and sick animals. Furthermore, in nature, very few male animals are required for reproduction and a large percentage of them falls victim to savage fights among themselves before reaching full adulthood. The cliché of 'survival of the fittest' perfectly sums it up and slaughter is therefore, unfortunately, an unavoidable part of livestock life.

People in Rajasthan had a traditional taboo on using camels for meat and, until the very end of the twentieth century, would never have considered slaughtering them. This worked fine as long as male camels were an important source of transport. But when they were replaced by cars, trucks and tractors in the 1990s and early 2000s, there was no longer a use for them. A clandestine trade in camels for meat developed. Although nobody talked about it and those in the know refused to admit it, camels were smuggled across India's borders to its meat-hungry neighbours in Bangladesh and Pakistan. Others were taken to cities in South India such as Bangalore and Chennai for sacrifice during Muslim festivals. Not a pretty idea to have camels walk over such long distances only to be slaughtered at the end. But at least it created income, with meat prices being high. There was still an economic incentive to breed camels, so the traditional camel culture teetered on.

Nevertheless, Rajasthan's camel population dwindled because of a lack of grazing areas and the spread of disease. Animal activists used the diminishing numbers of camels to convince the state government that slaughter was to blame for this. In 2014, a law was passed that prohibited the slaughter of camels and even their movement across state borders without a special permit. The animal protectionists celebrated this as a huge victory. But the law had the opposite effect than the one intended. It threw the camel breeders into despair as it became virtually impossible to sell male camels or to earn any income from camel breeding. How could they afford to spend all their time herding and breeding camels if it did not generate any income for them? Out of desperation, many of them left their camels to roam around by themselves. Deprived of healthcare, especially for mange, a highly contagious skin disease, many of them died. Male camels roaming around in the desert turned dangerous, starting to attack humans when they came too close to them. Until then this was unheard of in Rajasthan, although it is a familiar situation in Australia where camels have been feral for generations. There they are regarded as extremely dangerous and people are advised to immediately shoot them when they encounter them, as Robyn Davidson describes in her book *Tracks*.

Other camels were sold 'illegally' to traders who smuggled them across the Rajasthan border and took them to slaughterhouses in Hydrabad or Meerut. On the way, the trucks carrying them are regularly intercepted and confiscated by animal activists who then raise funds for transporting them back to Rajasthan 'where they belong', but where nobody wants them. They once again get loaded into a truck that takes them to a kind of camel sanctuary in the Sirohi district of Rajasthan that is chronically underfunded. The road is long and the journey takes several days, the camels regularly get injured and all of them are deeply traumatised. Some of them are 'adopted', only to often end up again on a truck to the abattoir.

The best strategy from an animal welfare perspective would be to have an expertly designed abattoir in Rajasthan where male camels could be humanely slaughtered without having to be trucked anywhere. (Even better, of course, would be finding a use for them as a riding or cart animal, but this seems to be a long way off, although climate-wise it would certainly make sense.) Sadly, the current political climate does not allow for the slaughter of camels to even be discussed, but at the time of writing the law is being revised to allow crossing of state borders.

---

Slaughtering an animal does not have to be cruel. It must be quick and happen in the normal surroundings of the animal without the stress of being transported to an unfamiliar place. While I was doing my ethnoarchaeological research among the Amareen Bedouin in Jordan, I had the opportunity to witness just this.

I was put to shame when my host Musa gently admonished my own three-year-old toddlers that had been piling onto one of the family's donkeys to treat it with more respect. Whenever I returned for a visit, often bringing along friends, they would treat me to Jordan's national dish *mensaf*, as part of their traditional hospitality ritual. *Mensaf* traditionally consists of a big plate heaped with flat bread or rice, piled onto which is a whole chopped-up sheep or goat that has been cooked in a broth with ground jamid (a kind of dried yoghurt). The Bedouin usually served a somewhat watered-down version of the real *mensaf*: they would buy chicken in Wadi Musa and then prepare it mensaf-style. But one time when my mother was accompanying me, they decided this warranted the real thing and was worthy of sacrificing one of their male goat kids.

Musa caught one of the goat kids, which was maybe four months old, carried him to a rock and held him down. Then his brother-in-law, Ibrahim, came with a long knife that he must have sharpened, quickly cut through the aorta and the blood streamed out. The kid hardly whimpered, nor did it struggle; it seemed as if its blood and its life drained away. It was so peaceful that even squeamish me could watch it without averting my eyes.

Once the blood had stopped running and the body was lifeless, it was strung up by the hind legs to the branch of a tree. Ibrahim and Musa jointly skinned it and subsequently dismembered the carcass, then threw the parts into a big cooking pot which they handed over to the women in the separate, non-public part of the tent.

It was all over in half an hour.

---

Pastoralists rarely slaughter for their own consumption; eating meat is reserved for special occasions and celebrations. Instead, most of them produce for the global market in beef, mutton and camel meat. Although they treat their animals well as long as they are in their care, they generally close their eyes to what happens to the animals once they have sold them.

Live animals produced by pastoralists are routinely trucked over hundreds of kilometres or transported in special ships between continents. Ethiopia and Somalia supply the Arab world, and thousands of camels are routinely walked up from Sudan to Cairo on the famous 40-day road to meet their end in a slaughterhouse where conditions are described as horrendous. Live sheep are shipped from Australia to the Middle East and cattle to Indonesia and other countries. Among some Australian 'pastoralists', or rather ranchers, awareness about the need to take full responsibility for the whole life cycle of their animals is building. In 2011, live cattle exports from Australia to Indonesia were banned by the government after animal-welfare groups exposed the cruel conditions under which the animals were slaughtered. This led to enormous losses for many producers, but Charles Massey, a prominent sheep breeder and author of a book on ecological farming, took a stance by saying that breeders cannot dodge the moral issue or abdicate that responsibility and must take responsibility for the welfare of animals until the end. I am afraid that this issue is often not on the radar of the traditional pastoralists who are the subject of this book and are not yet concerned about the issue.

While the situation will rarely be as ideal as I described for the Bedouin, in rural India, and even in Africa, slaughter of sheep and goats in small towns involves minimal cruelty, whether the animal is killed according to Muslim *halal* practices or by the *jatka* method prescribed for Hindus. The animals are tied up in a yard or outside the meat shop and provided forage to nibble on. They are slaughtered in intervals – when all the meat of the previous animal has been sold. The actual act of slaughter happens so fast that they hardly know what is going on.

In big slaughterhouses, the situation is entirely different. Unfortunately, the trend is going towards eliminating small abattoirs, allegedly for the sake of efficiency and hygiene. The regulations in the EU have led to the gradual disappearance of small slaughterhouses in the countryside. In the U.K., there is a campaign by the Sustainable Food Trust to bring them back, so that animals not only have a good life but also a good death.[2]

There are also efforts led by Temple Grandin to improve slaughterhouses so that they do not intimidate and cause stress to animals. Temple has been commissioned by some of the big US chains, such as McDonalds, to audit abattoirs and make them more animal-friendly. She has suggested changes that avoid causing stress, such as minimising noise and installing non-slip floors.

The gold standard for humane slaughter is to kill the animal on the pasture by shooting it from a short distance. The animal does not experience any stress and is killed instantly. It has positive effects on meat quality in terms of tenderness, colour and water retention.

The emergence of 'ethical butchers' in both the U.K. and United States is another positive trend. Many of them are former vegans or vegetarians who have opened shops that sell meat from animals raised naturally and without wasting any part of the animal. The larger goal of this movement is to provide an alternative to impersonal industrial food production and re-establish connections between animal, farmer, butcher and cook.

I want to give the final word to a friend of mine, Ruth Häckh, who is one of the last nomadic shepherds in Germany and has a small slaughter facility. In her book, *One for All: My Life as a Shepherdess*, she describes growing up with animals being slaughtered and this being a normal part of shepherding. She feels it is her responsibility to slaughter her animals herself because that is the only way she can ensure they do not feel frightened and that their good life is followed by a good death.

PART THREE

# Why Herding Is the Future

# Conclusion

*The animals of the world exist for their own reasons. They were not made for humans any more than black people were made for white, or women created for men.*

ALICE WALKER[1]

Pastoral systems have been around for around 9,000 years, at least a hundred times longer than the intensive systems that started to be developed in the first half of the twentieth century. Herders have experience in producing food reliably even under highly variable and uncertain conditions. If we want to get animal farming back on track again, we must take inspiration and learn from them. We need to understand them better and reflect on which aspects of their way of managing livestock we can integrate into livestock production systems in the 'North' to make them more ecological and animal-friendly.

Our problems with livestock started when we began to treat animals as if they are plants. Nature designed animals to move, and movement is what distinguishes them from plants. Plants can directly convert sunlight into energy by means of photosynthesis. But animals lack that facility. If they stay put, they exhaust resources and die. If they want to live and reproduce, they have to move TO the plants to obtain energy. Furthermore, from the perspective of the planet, the role of animals is to dissipate the energy that plants keep accumulating (thereby preventing fires) as well as to return carbon and nutrients to the soil, so the plant cycle can start anew.

Plant-eating animals – herbivores – have the planetary function of recycling nutrients – either cycling them down into the soil to feed micro-organisms that help plants to grow or upcycling them into their body mass to feed predators higher up the food chain.

## Where it went wrong

Over the last century, we have meddled with that basic principle of planetary design and turned it upside down. Through our human pursuits, especially agriculture that has wiped out their habitats, we have reduced the planet's wild fauna by more than 90 per cent. Alongside this extinction, we have exponentially increased the number of farm animals, confining and stall-feeding them instead of letting them graze or browse on natural plant material and fulfil their ecological functions.

In order to run this system and to feed more animals in cramped, factory-like surroundings, it has become routine to grow feeds on one side of the world (Brazil, USA and Argentina) and then move them to the other side of the world (Europe, Southeast Asia and China), although the pollution caused by just one container ship is said to be equivalent to that of 50 million cars![2] This system, controlled by a handful of corporations, requires an enormous expenditure of energy in the form of fossil fuels for cultivation, fertilisation, harvesting, transportation and many other intermediate steps in provisioning and arranging animal feed. It entails a massive destruction of biodiversity and the release of toxic chemicals into the environment, quite apart from treating animals as inanimate objects and not allowing them to be social beings.

Since Robert Bakewell (1725–1795), the putative father of modern animal husbandry in the eighteenth century, doubled the meat production of his herds through a combination of breeding and feeding, thereby providing a model for other breeders who wanted to improve livestock to cater for the demands of a rapidly growing and industrialising population, for almost a hundred years, scientists have focused their energies on creating breeds of animals that perform best in such artificial systems – but that can't survive outside them. In the process, we have reconfigured livestock from intelligent, solar-powered harvesters of biodiverse vegetation into guzzlers of human-edible protein and carbohydrates and, indirectly of fossil fuels, without considering what this is doing to their mental well-being and quality of life as well as to the environment and planetary functioning. This development has been driven by commercial interests that earn from each of the inputs – genetics companies, fertiliser companies, feed companies, veterinary companies – and steps along the supply chain.

Unfathomably, this approach has been validated by livestock scientists, especially the economists among them, as 'efficient', based on the yardstick of feed input versus food output and the amount of Greenhouse Gas (GHG) emissions per kilogram of milk or meat. Their endorsement of this destructive approach has justified the corporate sector gaining almost complete control over our food systems, while seducing us with low prices and convenience foods designed to be addictive.

This model of generating animal-sourced food is not only hugely damaging, but also unstable and risky, as it is based on monocultures of both plants and animals. Monocultures are prone to diseases and can only be upheld with the routine use of pesticides in the case of plants and of antibiotics in the case of animals. Growing animals in confinement, as dictated by 'efficiency', brings additional risks. It is unhealthy for body and mind and creates ideal conditions for the emergence and transmission of diseases. It also depends on perfectly functioning global supply chains, which are no longer a given in these unstable times of wars and pandemics.

## Integrating nature and food production

We regard agriculture and nature conservation as two independent domains handled by separate ministries and institutions. So, on the one hand, in the name of food security and increasing yields, we carry on with agricultural practices that kill biodiversity, denude the soil and heat the globe. On the other hand, to compensate for these wrongs, we espouse conservation and seek to preserve nature in what we believe is an inviolate state. However, it is in pastoralist areas that nature is often best preserved, and these areas host a panoply of wildlife exactly because of traditional herding practices.[3] Ironically, the thrust to conserve nature frequently leads to the eviction of pastoralists from their customary grazing areas in various countries, ranging from China to India to Tanzania and elsewhere. It is nonsensical to intensify agriculture while destroying those systems that produce the most nutritious food in the most natural way. Our goal must be to integrate food production and nature conservation, instead of going to war with nature here and erecting fortresses for her protection there. Fortresses always fall eventually if they have no hinterland. So, we need nature-based solutions to food production and the future lies in an integrated approach, in coaxing nature and arranging her cycles so that she can feed us all.

From the perspectives of both energy use and epidemiology, the mainstream approach to livestock production, which continues to be promoted even now, is a dead end from which we need to find an escape route as rapidly as possible.

To achieve this necessary turnaround, we need to imagine livestock differently and take our cues from the original ecological function of animals. We must acknowledge that the Earth's resources are limited and must be used wisely: our focus has to be on enabling livestock to make optimal use of the plant resources that the Earth provides naturally or that are available as by-products of crop cultivation. The main rule must be to minimise fossil fuel use for producing animal feed. In addition, we must make the most of the potential of livestock to sequester carbon and recycle nutrients into the soil, as nature intended animals to do.

In order to help livestock achieve this, we need to apply the principles of diversity and dispersal to livestock keeping. We require a diversity of animals that fit varied ecosystems, are robust and resistant to the spread of diseases, and can walk to faraway areas to utilise the seasonal vegetation there. Instead of confining and concentrating them in feedlots, we need to disperse livestock widely, in tune with the availability of local feed resources. In agrarian areas, livestock systems must again become circular and be re-integrated with crops so that nutrients are recycled and manure once again turned into an asset that upholds soil fertility, rather than a toxic mass that pollutes groundwater and water bodies. From the nutritional perspective, we need animal diets rich in phytochemical substances, for both their health and ours. On a general level, we need flexible systems that can respond to changing circumstances.

## What pastoralists can teach us

Humanity's herding heritage is a fount of wisdom for achieving food production that is in tune with nature and treats the Earth as she deserves. Unfortunately, the wealth of applied pastoralist knowledge has been maligned and suppressed since colonial times when an agrarian worldview originating in the temperate climes of Europe was imposed on countries in the south. Pastoralism has been demonised by 'regimes' founded on sedentary farming who believe they can control and shape nature according to their whim. Herding keeps being ignored and belittled by a reductionist scientific paradigm that conceives livestock as input-output machines in an ecological vacuum and ignores its interactions and connectedness with all realms of the natural world – soil, water, air, biodiversity, disease.

One of the treasures bestowed on the world by pastoralists are livestock breeds that on the one hand retain characteristics of their wild ancestors by being tough, resilient and able to withstand extreme weather and fluctuations in forage availability, but on the other hand are also able to respond to human directions and have a relationship of trust. These breeds are the future, not the genetically uniform, high-yielding animals whose metabolisms are only oriented at production. At the danger of sounding utilitarian, herding cultures have developed the best tools humanity has for adapting to and coping with a changing climate.

The other innate trait of herding cultures is their flexibility in adapting to changing circumstances: they move their animals here and there, adapt animal numbers and types, shift family members in and out of herd management chores, delay animal pregnancies, and slaughter according to what circumstances demand. They can react quickly, contrary to animal industries in which operations are tightly timed and fall apart when one factor changes. Risk-management expert Emery Roe from the Center for Catastrophic Risk Management (CCRM) at the University of California, in Berkeley, has dubbed pastoralists 'reliability professionals'.[4] He regards them as providers of critical global infrastructure and as a system relevant for the planet. And they are. Even if everything, including the long-distance supply of energy and the internet, breaks down, pastoralists can still function by relying on biological processes that go on regardless of all such disturbances.

## Changing the paradigm

In order to make policy makers appreciate the multitude of benefits and advantages of herding, we need to de-colonise animal science. Its current fixation on efficiency and focus on greenhouse-gas emissions in relationship to milk or meat obscures the fact that sustainability also encompasses biodiversity, soil health, air and water quality, public health and rural livelihoods. It is therefore a matter of urgency that animal science adopts a more comprehensive way of looking at livestock production systems and takes these aspects into account. Only then will policy makers understand and properly acknowledge the value of the nomadic pigs in Odisha that roam around in happy family packs, upcycle crop by-products into meat and downcycle them into organic fertiliser, while feeding the poor and providing livelihoods for the marginalised.

But there are big hurdles for the animal science establishment to overcome if they are to revisit its foundations and develop a more holistic framework in tune with changing times and the demands of the public. Too many commercial interests are invested in the present system – the feed companies, the fertiliser and pesticide conglomerates, the genetics companies and the veterinary drug corporations.

## What needs to happen

So how can we revive pastoralism? The most urgent task is to secure long-term access to grazing areas by recognising and formalising pastoralists' customary rights. The space for herding has been shrinking and continues to do so. Throughout much of the world, the combination of population pressure, urban sprawl, road construction, irrigation projects and wildlife conservation areas is squeezing pastoralists out of the landscape. There is simply no room to move with animals; former grazing areas have been fenced off, built upon or declared inviolable for the purpose of nature conservation.

In addition, land use plans have to reserve space for the movements of pastoralists even in the face of 'development' such as highways, urban sprawl, industrial areas, and so on. If we can create corridors for elephants in India or for frogs in Europe, we can also do the same for pastoralists. Maybe we should take inspiration from the Mesta, the livestock keepers association in Spain that put in place the network of *Cañadas* that criss-crosses the country for the sake of keeping sheep and other livestock healthy and productive by ensuring good pastures. As we have seen, this also does wonders for biodiversity.

Even if it is physically possible, as it is in quite a few remote areas, young people from pastoralist families are frequently not attracted to a profession that entails hard work around the clock and constant exposure to the elements. The widespread perception of a herder as being dirty, backward, poor and uneducated is another disincentive. Even if herding provides better income opportunities than menial labour in cities, it is not a desirable way of life. Life in the city beckons.

At national policy levels, pastoralists remain unseen, if not unwanted, as a category; nobody really knows how many of them there are and no country has a plan for them. Mongolia is the only nation that is supportive of herding and regards pastoralism as part of its national identity, since crop cultivation has never played a role in its economy. But there, too, herding is on the decline, as mining is also a key economic sector and proceeds at the expense of pastoral land use. Young women, traditionally the mainstay of pastoralist families, have become well educated and many of them opt for urban careers. We need to encourage young people to take up the profession. We can only achieve that by making pastoralism an attractive modern proposition that is no longer regarded as backward. Pastoralism combines biodiversity conservation with food production, so it must be turned into a desirable career option with high status and a decent income. Pastoralists should be the new rangers, the avant-garde conservationists!

Yet global awareness about pastoralists is undeniably growing and, in 2015, the Food and Agriculture Organization of the United Nations (FAO) set up a 'Pastoral Knowledge Hub' that undertakes various activities in data collection and raising awareness. In 2021, it officially stated its intention to 'mainstream pastoralism' within the organisation and published a detailed report on pastoralism that highlights its advantages in working with nature rather than against it.[5]

While there are thus promising moves at UN level, it takes a very long time for their impacts to trickle down to the grassroots level through the various bureaucratic levels, encountering resistance by other interest groups and the deeply ingrained thought model that productivity is everything.

This process is much slower than the speed at which herding knowledge, grazing areas and pastoralist livelihood options are being lost. Therefore, ground-level initiatives are of much more practical relevance. They achieve impact in the here and now and are leading the way towards more natural ways of raising livestock.

## Developments in the Global North

In those parts of the world – the 'North' – where confined animals are the norm and default model, and where traditional herding systems no longer exist, we need to encourage more scope for moving livestock through policies and incentives.

In the last decade or so, many initiatives have sprung up that emulate key aspects of pastoralist land and herd management. The provocative rangeland guru Allan Savory, who developed the concept of Holistic Resource Management (HRM) in order to restore local ecologies by means of grazing, has spawned an extensive movement and global network of practitioners and supporters (see 'Regenerating Land', page 143). Holistic management has been taught to more than 50,000 people in over 130 countries, according to the organisation's website.[6] The frequent movement of animals and constant monitoring of range conditions are features that Holistic Resource Management shares with pastoralism.

That things can be turned around when dedicated individuals devote a lifetime of work and passion to an ideal is demonstrated by the achievements of Jesús Garzón-Heydt (widely known as Suso). In 1992, transhumance had been officially declared extinct in Spain. Suso's enthusiasm infected many others, both on a practical and academic level; he was invited to give speeches at universities, inspired numerous PhD students whose research proved the benefits of transhumance for biodiversity, was taken seriously by top ecologists and written about by well-known rangeland scientists. He also had a way of skillfully handling the media and when, in 1994, sheep flocks once again walked through Madrid after a hundred-year interval, this was picked up by international news channels. In effect, he has revived a tradition that had been forgotten and which only a few old shepherds remembered. The Spanish transhumance has become famous and known all over the world.

The Spanish experiment has set an example and inspired others to follow, in Europe and beyond. Working together with Suso is Francesca Pasetti, an Italian expert in ecology, who worked in vulture conservation in Mallorca and became well aware of the link between their survival and mobile livestock keeping. Together with Nicola di Niro, a rural development expert, she is one of the coordinators of an initiative that aims to put Europe's largely forgotten herding cultures and drove roads into the centre of rural development. This visionary idea is called *Programma TRE (Terre Rurali d'Europa)* and originated in South Italy where transhumance also has a long history, as it does in the whole Mediterranean area and many other countries. Here, the drove roads are called *tratturi* and connect winter grazing areas along the coast with summer pastures in the mountains. About 100m (300ft) wide, with an average length of 200km (120 miles), and stretching over a total of 3,000km (1,850 miles), they go back to pre-Roman times and are associated with a rich cultural heritage. According to the organisers of TRE, such drove roads once connected Constantinople in Turkey with Finisterre in Spain, and Sicily with the Arctic Circle, and were the paths along which civilisation spread. The TRE programme aims to use the drove roads as nuclei for integrated rural development, reviving local historical, cultural and environmental heritage, as well as regional food products and crafts, and stimulating rural development with territorial identity throughout Europe, thus stopping rural-urban migration.

The activities around reviving traditional drove roads have raised awareness that transhumance was once a Pan-European phenomenon present from the far North to the far South. In 2019, transhumance was recognised as UNESCO Intangible Cultural Heritage in Italy, Greece and Austria. Currently, seven more countries – Albania, Andorra, Croatia, France, Luxemborg, Spain and Romania – are preparing applications for the same status. The fact that something which is regarded as outdated and behind the times in their countries has achieved this recognition has sent a message to non-European nations, inspiring the pro-pastoralist groups in India and Mongolia, as well as many other Asiatic and Latin American countries. A third application by non-European countries such as India and Mongolia is now being discussed by the promoters of UNESCO recognition.

In order to teach the skills and knowledge required to successfully lead a flock across the landscape, a string of public and privately supported shepherding schools have opened in France and Spain where young people can learn the art and science of shepherding. These in turn have inspired nature enthusiasts from the US to set up similar programmes there, especially on the West Coast. Thus, a new generation of herders has been born in the Global North: ecologically minded young people take up herding to restore land, conserve biodiversity, counter climate change and prevent forest fires. They have no family background in herding, but they want to contribute to planetary health in a practical way. They rent themselves – and often their flocks of sheep and herds of goats – out to private landowners or public institutions to manage their properties in an ecologically sustainable way.

In both the US and Europe, there are now many initiatives to promote herding as a way of restoring land and soils as well as producing extremely high-value livestock products. In the US, the Pasture Project of the Wallace Center, which supports communities to regenerate ecosystems, has set up a match-making programme for graziers and landowners, so that herders can access forage and landowners can find herds to improve their land and its soil.

In the U.K., the Soil Association established the Pasture-Fed Livestock Association (PFLA), which supports regenerative grazing and includes farmers, butchers, academics and consumers and awards a certificate for meat and dairy from pasture-raised animals. The Shieling Project in Scotland seeks to familiarise the public and especially children with this ancient practice.

## Revival in the Global South

In the Global South, the situation is different. It is not so much a question of reviving or going back to herding, but of supporting and strengthening what is still there. In large parts of Africa and South Asia, herding is very much alive and often the dominant way of livestock keeping, making major contributions to national economies – between 10 and 44 per cent of GDP in African countries – and generating significant amounts of foreign currency.[7] For instance, in the Horn of Africa, camel pastoralism plays a huge economic role, both at the subsistence and export level, with large numbers of camels being exported to Arab countries for meat, while the local trade in camel milk is in the hands of women entrepreneurs. In India, agro-pastoralism remains the backbone of the livestock economy, producing more meat and milk than the stall-fed systems.

The problem in Africa and South Asia is that policy makers and livestock experts, trained in formal animal science (often in the US or Europe), do not recognise the assets that their herding systems represent and either ignore them or do nothing to support and nurture them. They aspire to copy the 'efficient' livestock systems of the 'North', instead of focusing on protecting grazing areas and migratory routes from land use change and thereby ensuring that herders are not squeezed out.

A number of vocal NGOs and the World Alliance of Mobile Indigenous Peoples (WAMIP) have been raising these issues for two decades and some inroads are being made, although too slowly. But herders all around the world are grabbing the opportunity to speak up on the global stage, especially in the context of various international legal frameworks that protect indigenous peoples and the conservation of biological diversity.

In the first decade of the third millennium, advocacy by and for pastoralists really took off.[8] Galvanised by Uncle Sayyaad's plea at the World Parks Congress in 2003 to make pastoralists equal partners in conservation, international agencies competed to provide them with a platform.[9] Suddenly, articulate pastoralists were much in demand and it sometimes felt as if we were running a speaker's bureau plus travel agency for Raika representatives (and their translators) to attend meetings in far-flung places, ranging from the UN headquarters in New York to remote pastoralist areas in the south of Ethiopia, a national park in Mongolia and the annual passage of sheep flocks through central Madrid.

In this overall context of pastoralist advocacy activities, one particular process revolved around the concept of 'Livestock Keepers' Rights' when the conservation of livestock biodiversity was intensely discussed at UN level. In 2003, some 40 leaders of pastoral communities from across Africa, India and Mongolia and an additional 20 participants from non-government, government and research institutions came together in a workshop in Kenya to point out that it was pastoralists, rather than scientists, who had created and continued to steward much of the world's livestock biodiversity by means of their indigenous breeding practices. A milestone in pastoralist advocacy, paraphrased, their message was this:

We, herders and other livestock keepers, are the ones who have created and are stewarding the world's diversity of livestock breeds and we want that to be recognised! We will continue doing so, if you afford us the opportunity by securing the areas in which we developed these breeds and by recognising the importance of our traditional knowledge. We need to have the freedom to make our own breeding decisions ourselves and not be stopped from that by corporations, we need to be involved and heard when livestock policies are made, we need help with building value chains, so the products of our breeds generate income, and we have the right to information.[10]

In a series of further meetings, pastoralist leaders eventually identified a bundle of rights or privileges that they would require to stay in business: land access rights, respect for their knowledge, participation in policy processes and support for building up local value chains. The aim of this bundle was to protect the rights of pastoralists and other small-scale livestock keepers over one of their core activities: making their own breeding decisions without interference by corporations (and the associated danger of appropriation) and thereby stewarding diversity and the kind of resilient livestock that we need for the future.

In the current climate where there is so much antagonism against livestock – with some people wanting to remove it from the planet completely – there is certainly a renewed relevance to the concept of 'Livestock Keepers' Rights' to ensure that the ecologically positive livestock keepers – that is, the herders – survive, by addressing the needs they themselves have identified. Furthermore, they may be especially relevant in the future to support 'good livestock keepers' who keep animals as part of local ecologies rather than in industrial production systems.

## The need to unite herders of the South and the North

For herders in the South, herding is a hereditary occupation that they have taken over from their parents and grandparents and uncountable generations before. Living in close symbiosis with their inherited herd is the only way they know of looking after animals, and they would be aghast to see what passes as 'modern animal production'. By contrast, many, although not all, the graziers in the North are first-generation herders motivated by the wish to conserve and restore ecosystems, to produce food in a sustainable way and to do something meaningful with their lives, rather than serve corporate interests. Regardless of their backgrounds and whether they are located in the South or in the North, herders must urgently come together on a common platform and unanimously advocate for the recognition of herding as a globally relevant way of addressing climate change, biodiversity loss and land degradation.

The recent successful campaign of the Government of Mongolia to convince the United Nations General Assembly to declare 2026 as the International Year of Rangelands and Pastoralists (IYRP) has created just such a platform. More than 100 governments and 303 organisations have joined the campaign, which has spawned about a dozen regional support groups that are already preparing activities for 2026.

The decision of the FAO to mainstream pastoralism, the panic caused by the pandemic about future disease threats, and the geopolitical situation that is leading to a shortage of chemical fertiliser, grain and energy are all forces that push us to move away from corporate-controlled industrial animal production – and to espouse livestock keeping in tune with nature, as exemplified by herding.

The events of the past three years have shown the limits of globalised livestock value chains and underscore the need for investment in networks of decentralised livestock processing facilities. While there is an ecological cap on the number of pastoralists as primary producers, there is huge scope in secondary value addition to livestock products. This is where profit lies. We need to support herders' associations with grants and soft loans and mentor them in the marketing of pastoralist products as eco- and animal-welfare friendly.

## The place of animals – as part of humanity

Finally, there is much more to our relationship with farm animals than ecology, livelihoods and nutrition. Livestock are our link with the Earth and by being close to them we grasp something that is bigger than us: the eternal cycle of life that is composed of birth, play, work, reproduction and death. This cyclical experience cures us of the human hubris of being the masters of the Earth and reminds us that we are mortals, too, links in a chain that stretches way back from our ancestors to our descendants far in the future (hopefully). A planet without livestock would not only be terribly impoverished, it would also hardly be alive, taking us one step closer to becoming robots ourselves. A world without farm animals would deprive us of our humanity.

Many children and young people yearn for that connection, and it makes them better people. The Bergers Urbains, young urban herders with no family background in herding, ostensibly deploy flocks of sheep to graze the green spaces of Paris, but what is more important is that they give city kids the opportunity to familiarise themselves with farm animals.[11] We must not lose that connection. Livestock are not an obsolete technology that one chucks out because something newer and better is available, as advocated by some false prophets that want to take us into an even more artificial future. The bonds that we humans have developed with selected non-human animals over the last ten thousand years have taken us to the ends of the Earth and been an integral part of our cultural journey. Yes, we in the Western world have been led astray by applying efficiency thinking and profit motives to this relationship, and it has become exploitative and toxic. But that does not mean we should abandon it. Instead, we must invest in getting it back on track. Restore it to a mutually supportive basis and apply our joint intelligence to getting us and the Earth off our current destructive trajectory. When you are in trouble, you do not abandon long-standing relationships; instead, you nurture them back to health. This is the task before us, and the wisdom of our overlooked nomadic heritage can help us do so.

---

Herding is therapy, not just for the planet, but also for our souls.

# Epilogue

In the Godwar area, camel numbers have once again been increasing. For a while it looked as if camel herds would totally disappear from the local landscape. Ironically, the situation reached its lowest point after the camel had been declared Rajasthan's state animal in 2014 and a law was passed in 2015 that aimed to save the camel by prohibiting its slaughter and export across state borders. At that point in time, we decided to set up the Kumbhalgarh Camel Dairy as a means of creating income for the camel herders. It was not easy. But we managed, raising the funds to construct a micro-dairy and buy freezers, and taking out a loan to afford marketing expertise. It is a micro-enterprise, struggling to be economically viable. But it has achieved some success: around a dozen Raika families make a good income, camels have returned to the landscape, autistic children and cancer patients benefit from the healing qualities of the camel milk, visitors from near and far can catch a glimpse of the Raika camel culture as it once was, and local schoolchildren have the opportunity to learn about camels. We even have an agreement for providing 'cruelty-free' milk to a start-up that produces healthy foods, but has animal welfare as a core value.

But now another threat has raised its ugly head: The Kumbhalgarh Wildlife Sanctuary, the indispensable rainy season grazing grounds for Raika camels, is to be converted into a Tiger Reserve. The existing reserves in Rajasthan are too small for the growing number of tigers. Although experts deem the Kumbhalgarh area as not suitable for tigers, the forest department is forging ahead regardless. It has begun building a huge wall and set up a 'Herbivore Enrichment Centre' in which gazelles and antelopes are bred to serve as prey for the tigers. Politicians and other powerful groups are constructing hotels around the boundary of the planned reserve to benefit from the expected stream of national and international tourists. Tribal people living in the reserve have been sent notices to vacate their settlements.

A fortress for tiger sightseeing is being created – an enlarged zoo centred on only one species. A design without the local people living in and around the sanctuary. A construct that ignores other wildlife that is even more threatened, such as the wolf and the striped hyena. A plan that means the death knell for local camel herding. A programme that will eliminate the Raika shepherds and goat keepers who I meet on my daily walk and who have been such a big inspiration behind this book. It is a plot that sees humans as separate from, and incompatible with, nature. It is not a solution.

# Acknowledgements

It feels as if almost the whole herding universe has midwived this book. My fascination with herding goes back to 1979 when, while working on an archaeological site in Jordan, I made the acquaintance of a Bedouin family who sang songs to its camels and was happy with a minimum of posessions. The goat herding Bedouin family I stayed with while conducting an ethnoarchaeological survey around Beidha in southern Jordan in 1982 further deepened my appreciation. After a family break, I had the opportunity to survey the Rashaida camel pastoralists around Kassala in eastern Sudan, thanks to Professor Bakri Musa, then the dean of the Veterinary Faculty at the University of Khartoum, and to the Agnese Lindley Foundation.

But my engagement only began in earnest in 1991 when I had a fellowship from the American Institute of Indian Studies to research camel management in Rajasthan during which I was exposed to the marvels of Raika camel culture and became aware of the threat it was under. I am immensely indebted to Dr Dewaram Dewasi who provided me with an introduction to the community and educated me about its mores. That is when the unexpected journey that led to this book really started.

Subsequent fellowships by the Alexander von Humboldt Foundation and the German Research Council allowed me to study the history and ethnography of global camel pastoralism in depth, while early support from Misereor in Germany made it possible to combine this with veterinary support for the Raika camels.

For quite some time, it seemed as if the Raika culture was an isolated phenomenon, but gradually a network of NGOs working with pastoralists in India coalesced, and since the early 2000s, it has become evident that India is choc-a-block with herding communities similar to the Raika in their ethics and ecological roles. I feel very close and will always be deeply grateful to Balaram Sahu, Kamal Kishore, Gopikrishna, Vivekanandan, Chanda Numbkar, Nitya Ghotge, Bhavana Rao, Aman Singh, Ramesh Bhatti, Selvarajan Rajeshwaran, Nilkanth Mama, Karthikeya Sivasenapatty, and many others who came together in the LIFE Network for community-based conservation of local livestock breeds to promote the recognition of pastoralism in India. Without you, I would have never been able to get so many insights into Indian pastoralism!

Equally important were and are my close friends who came together to form the League for Pastoral Peoples (LPP) in 1992, initially just to provide veterinary support to Raika camels, but eventually embarking on the endeavour of pastoralist advocacy in general and getting the world to understand the importance of herding. Christiane Herweg, Juliane Bräunig, Bettina Bock, Bruno Haas, Aisha Liebehenz: you have been with it from the first hour and never wavered in your trust and support! Günther Czerkus, Silke Brehm and Micky Wiesner joined later, but have been pillars of support. This book would not be here without you!

Special mention is due to Sabine Poth who was not there at the beginning, but joined us more than 15 years ago and, since then, has been the heart and soul of the League for Pastoral Peoples, who has kept it going through thick and thin, the most conscientious, diligent and enthusiastic administrator one can imagine, gradually turning into an expert on pastoralism in her own right.

Crucial in developing the arguments in this book have been the many researchers and advocates of pastoralism; it has been a privilege and joy to work with: Abdul Raziq Kakar, Saverio Krätli, Pablo Manzano and Maryam Niamir-Fuller, as well as the members of the international LIFE Network: Maria-Rosa Lanari, Jacob Wanyama,

Elizabeth Katushabe, Evelyn Mathias, Paul Mundy, and many others. (Forgive me if you are not mentioned.) Jesús Garzón-Heydt has been immensely inspiring, and I am grateful to him and Francesca Pasetti for providing me with a glimpse into Spanish transhumance. I thank fellow veterinarian Katy Gomez for making available her stunning photographs. I would also like to express my appreciation to the folks who have set up the Global Agenda for Sustainable Livestock (GASL), a multi-stakeholder platform for 'practice change towards sustainable livestock'. The principle of this platform is consensus, so the space for critical discourse is limited there, but it has helped me to understand the positions of livestock industry representatives, and I must laud its chairs and organisers – Henning Steinfeld, Fritz Schneider, Shirley Tarawali and Eduardo Arce-Diaz – for their fairness and willingness to give pastoralism space.

We have been fortunate to have funding partners that make LPP's work possible and none of them is more important than Misereor who has supported both our research and advocacy activities, as well as the local work with the Raika, uninterruptedly, and given me some space to work on this book. A special thank you is due to Sabine Dorlöcher-Sulser who promotes engagement with pastoralism within the organisation. The Ford Foundation in India, during the tenure of Vasant Saberwal, also supported a period of time to write on Indian pastoralism.

Many of my friends, including Liz Wedderburn, Anne Bruntse, Sabine Poth, Gary Rollefson, Kate Hardy, Heather Meurer, Christiane Herweg and Juliane Bräunig, have read and commented on the manuscript or parts of it. Big hugs!

Still, this book might have never seen the light of day if it wasn't for the lockdown and the tutoring of Kathryn Aalto. Having to undergo a lengthy quarantine when returning from India to Germany, motivated me to take one of Katy's online writing courses, which made all the difference and enabled me to write a query letter to Chelsea Green that caught the attention of Muna Reyal. Muna saw not only the potential of *Hoofprints on the Land* but also majorly contributed to its final shape. I am grateful to copyeditor Caroline West, Anne Sheasby, and Angela Boyle and the rest of the team at Chelsea Green for the diligence they applied to the text.

Once the book was reaching its final structure, Randy Jackson, Gloria Putnam of Angeles Crest Creamery, Sylvain Perdigon, Sam Harrison, Lani Malmberg, Sam Osborne, Herbert Nickel and Stephan van Vliet were kind enough to share their extensive knowledge in Zoom meetings, contributing important pieces of the argument.

One can't write a book without one's nearest and dearest fully supporting the endeavour. I thank my family in Germany: Gary, Aisha and Frederik, Jon and Konzy, Justus, Lotta and Jette – you have all been exceptionally understanding and supportive! Here in India, I am surrounded by a crew of people who are like family and make my life enjoyable, creating conditions for writing, as well as having something to write about: the Lokhit Pashu-Palak Sansthan (LPPS) family of Ramesh Bhatnagar, Shanta, Anjuman, Gulabi, Magan, and others. During the monsoon, there is the wonderful Madhuram Raika from whom I have learned so much and who is really the hero of this book. There are Dailibai Raika who massages my back when it suffers from too much sitting and the 'Khimji boys' who run the household. And it is all held together by my life partner Hanwant Singh Rathore, without whom none of the developments of the last thirty years would have happened.

This book is truly herd-sourced!

APPENDIX 1

# The Struggle for Recognition and Rights

For about a quarter of a century now, herders have been fighting for recognition and for their rights under existing UN conventions. All around the world, pastoralists are grabbing the opportunity to speak up on the global stage, especially in the context of various international legal frameworks, such as the United Nations Declaration on Indigenous Peoples (UNDRIP), the Convention on Biological Diversity (CBD), the Global Plan of Action on Animal Genetic Resources, the United Nations Convention to Combat Desertification (UNCCD) and the United Nations Framework Convention on Climate Change (UNFCC). These UN-level processes provide a platform for diplomatic and (mostly) civilised exchange between countries that do not see eye to eye. It is important that pastoralists now regularly have a seat at the table, so they can raise awareness about their existence and significance, although how far and fast UN resolutions are implemented on the ground and actually lead to positive change is another question altogether. Nevertheless, continued active participation is instrumental to moving forward and pastoralists have issued more than 15 formal declarations in the last twenty years.

Here is a summary of the major processes that took place and the points they made:

### World Herders Council

Probably the first ever initiative by pastoralists with global aspirations was the *Conseil Mondial des Éleveurs* or World Herders Council (WHC), which was founded in November 1996 in Dori, Burkina Faso, and originated among Fulani pastoralists. The WHC aimed to be an international association of breeders and searched for a new ethical framework for livestock keeping built on pastoralist heritage in the South and North, meeting annually at different locations. It had regional chapters in Sénégal, Burkina Faso, Niger, Mali, Cameroon, Chad, Ethiopia, Sudan, and in France, Germany and Switzerland. Due to its Fulani origin, it focused on cows, rather than all pastoralist animals.

World Herders Council, *Conseil Mondial des Éleveurs*

### Dana Declaration

The Dana Declaration was the outcome of a group of concerned social and natural scientists from all over the world meeting in the Wadi Dana Nature Reserve, in Jordan, on 3–7 April 2002, to discuss how to integrate mobile peoples into nature conservation. Their engagement led to the first speech by a pastoralist at the World Parks Congress, a global forum on protected areas organised by the International Union for the Conservation of Nature (IUCN), which was held in Durban, South Africa, in 2003.[1] In a follow-up meeting, the Dana +10 Workshop was held in April 2012, in which a number of pastoralist leaders participated, and the Dana +20 Workshop took place in September 2022.

Dana Declaration on Mobile Peoples and Conservation
http://danadeclaration.org/main_danaconference.shtml

### Livestock Keepers' Rights

'Livestock Keepers' Rights' is a concept and a grassroots movement that developed during the time when the Food and Agriculture

Organization of the United Nations (FAO) was concerned about the loss of livestock breed diversity in the first decade of the twenty-first century.[2] The FAO and governments wanted to counter the trend, but completely ignored the role of pastoralists and other indigenous livestock keepers in stewarding the resilient farm animal breeds that are so important for food security, especially while the climate changes and temperatures rise. The seed for the movement was laid at the World Food Summit held in Rome in 2002, just after 'Farmers' Rights' had been enshrined in the International Treaty on Plant Genetic Resources for Food and Agriculture. This legally binding agreement recognised the customary rights of farmers to save, use, exchange and sell farm-saved seed and propagated material, and to participate in decision making on issues related to crop genetic resources.[3]

Once the idea of 'Livestock Keepers' Rights' was born, it was fleshed out in a series of workshops in which a total of 300 pastoralists from 60 ethnic groups and 18 countries participated. A key event was a workshop in Kenya in 2003 which resulted in an appeal to governments and international organisations to recognise the role of pastoralists and other indigenous livestock keeping communities in upholding livestock genetic diversity.[4]

When governments met in Interlaken, Switzerland, in September 2007 to agree on and commit to a 'Global Plan of Action on Animal Genetic Resources', pastoralists and civil society members held a parallel meeting, where they worked out the Wilderswil Declaration on Livestock Diversity.[5] During the government meeting, the African block of countries made a valiant attempt to get Livestock Keepers' Rights into the final text outcome. However, developed countries regarded these rights as unchartered legal territory and therefore opposed them. In the end, the UN-sanctioned Global Plan of Action for Animal Genetic Resources did recognise and emphasise the important role of local and indigenous communities in upholding diversity. But it nevertheless casts animal genetic resource management as an issue that is to be controlled and managed by the state and gives livestock keepers only a supporting role, instead of making them the key actors.[6]

## UN Convention on Biological Diversity and Community Protocols

The UN Convention on Biological Diversity is an international legal framework that recognises the importance of the traditional knowledge of local and indigenous communities in stewarding biodiversity.[7] However, on a practical level, communities never received any benefits for their conservation efforts. In order to address this gap, in 2010, countries agreed to the 'Nagoya Protocol on Access to Genetic Resources and the Fair and Equitable Sharing of their Benefits'. This made the provision for countries to support 'Community Protocols', in which communities detail the genetic resources and traditional knowledge of which they are the custodians.[8]

This seemed a tailor-made solution for herders to claim their breeds as their own and make their role in conserving biodiversity visible. The Raika were one of the first groups to develop such a Biocultural Community Protocol (BCP) in which they set out their biocultural values and explained who they are:

We are indigenous nomadic pastoralists who have developed a variety of livestock breeds based on our traditional knowledge and have customarily grazed our camels, sheep, goats and cattle on communal lands and in forests. This means that our livelihoods and the survival of our particular breeds are based on access to forests, *gauchar* [village communal grazing lands] and *oran* [sacred groves attached to temples]. In turn, our animals help to conserve the biodiversity of the local ecosystems in which they graze and we provide assistance to the area's local communities. In this way, we see our indigenous pastoralist culture as both using and benefitting from the forests, in a virtuous cycle.[9]

The idea of community protocols caught on and NGOs started working with communities to put together their protocols. In 2012, Dailibai Raika travelled to Kenya to help with the development of a BCP by the Samburu pastoralists. The process was organised by a community member; however, it involved only a small number of Samburu and illustrated some of the challenges of developing community protocols. After working in the field, Dailibai was invited to share the Raika BCP at a conference of African indigenous leaders on community rights in Nairobi; in December, Dailibai travelled to a UN meeting in Montreal where she handed a copy of the BCP to a high-level official.

Unfortunately, after such a promising beginning, BCPs did not really take off. For one, developing genuine community-created protocols is a time-consuming and costly process that few funding agencies were willing to support. Secondly, some governments subverted the concept, thinking they could send out consultants to put together the documents in the name of a community, without real community involvement and participation. So, what is a truly brilliant concept for community empowerment is currently sitting on the shelf.

## World Alliance of Mobile Indigenous Peoples (WAMIP)

The World Alliance of Mobile Indigenous Peoples (WAMIP) was founded in Segovia, Spain, during a pastoralist meeting on 11–18 September 2007, in which 200 pastoralists from 50 different groups participated. The meeting, which was held in conjunction with a conference of the UN Convention to Combat Desertification, issued the Segovia Declaration of Nomadic and Transhumant Pastoralists.[10] They also participated in the annual crossing of Madrid by Spain's transhumant shepherds. According to its website, WAMIP is an independent, grassroots-movement, global alliance of nomadic peoples and pastoralist communities and their organisations. They represent customary forms of pastoralism and other forms of mobility as a livelihood strategy, while conserving biological diversity and using natural resources in a sustainable way. WAMIP facilitated the Global Gathering of Women Pastoralists in Mera, made inputs to several FAO livestock processes, and now has alliances in nine regions, including North, Central, East and West Africa; Central, South and West Asia; and Latin America and Europe. It is now preparing to take an active role during the International Year of Rangelands and Pastoralism (IYRP).

World Alliance of Mobile Indigenous Peoples
http://wamipglobal.com/

**International Year of Rangelands and Pastoralists (IYRP)**

On 15 March 2022, the UN General Assembly unanimously declared 2026 as the International Year of Rangelands and Pastoralists (IYRP).[11] This initiative was spearheaded by the government of Mongolia, but also driven by the enormous and strategic engagement of civil society all over the world. It has already succeeded in uniting herders from the South and the North on one common platform and has given rise to a host of regional support groups that undertake their own decentralised activities and raise the awareness of their governments about rangelands and pastoralism. Hopes are high that the IYRP will turn around the attitudes of governments towards their pastoralists in a major way!

International Year of Rangelands and Pastoralists 2026
https://iyrp.info

APPENDIX 2

# Declaration on Livestock Keepers' Rights

After governments had refused to include reference to 'Livestock Keepers' Rights' in the Global Plan of Action for Animal Genetic Resources that they agreed upon at the first International Conference on Animal Genetic Resources held in Interlaken, Switzerland, in 2007, there was obviously a sense of great disappointment and frustration among the coalition of organisations that had pushed for the recognition and rights of pastoralists. However, legal experts from the Global South emphasised that the requested bundle of 'rights' is already implicit in existing international legal frameworks. Hence, in a workshop with legal experts held in Kalk Bay, South Africa, in December 2008, the individual 'rights' were linked to international law and articulated in the standard legal language that is generally used in UN-level statements.[1]

---

**We, the LIFE Network and other organizations representing indigenous and ecological Livestock Keepers and supporting ecologically and socially sustainable livestock development,**

Affirming our commitments to international legal instruments relevant to Livestock Keepers' Rights, including the UN Convention on Biological Diversity, the United Nations Convention to Combat Desertification, the Global Plan of Action for Animal Genetic Resources and the Interlaken Declaration on Animal Genetic Resources,

Recalling the Universal Declaration of Human Rights, the International Covenant on Economic, Social and Cultural Rights, the United Nations Declaration on the Rights of Indigenous Peoples, the Convention on the Protection and Promotion of the Diversity of Cultural Expressions, the Convention (No. 169) concerning Indigenous and Tribal Peoples in Independent Countries, the Declaration on the Rights of Persons belonging to National or Ethnic, Religious and Linguistic Minorities, and other pertinent legal agreements,

Taking cognizance of the efforts and contributions of the LIFE Network, its supporting organisations and stakeholders in previous processes advocating Livestock Keepers' Rights, including especially the Sadri Declaration, the Karen Commitment, the Bellagio Brief, and the Addis Résumé,

Conscious of the important role of Livestock Keepers in the conservation and sustainable use of livestock diversity, climate change mitigation and adaptation, livelihood and food security, sustainable land use and management of natural resources,

Aware of the threats posed by industrial and corporate models of livestock breeding, which include privatization of genetic resources through intellectual property rights and the free market system,

Affirming that Livestock Keepers possess collective rights which are indispensable for their existence, well-being and integral development as custodians of local animal genetic resources,

Conscious that the livelihoods of Livestock Keepers are under threat by the progressive loss of grazing land, limitations to mobility and lack of participation in decision making,

We hereby recognize and endorse the following principles and rights:

**Principle 1:**
**Livestock Keepers are creators of breeds and custodians of animal genetic resources for food and agriculture.**

Over the course of history, pastoralists and small-scale farmers have managed and bred their livestock, exposed them to different environments, selected and used them, thus shaping them so they are well adapted to their environment and its extremes. Keeping these breeds is a vital part of their culture and livelihoods. Yet these breeds and their livelihoods are under risk through globalization, environmental degradation, climate change, changes in land use and many other threats. This might endanger food security in marginal areas.

As recognised in the Global Plan of Action for Animal Genetic Resources and the Interlaken Declaration on Animal Genetic Resources, livestock keeping communities are thus the creators and custodians of the breeds that they maintain. They have therefore earned certain custodianship rights over these breeds, including the right to decide how others use the genetic resources embodied in their breeds.

**Principle 2:**
**Livestock Keepers and the sustainable use of traditional breeds are dependent on the conservation of their respective ecosystems.**

Traditional breeds are developed through the interaction between livestock, Livestock Keepers and their natural environments. These natural environments are conserved, *inter alia,* through traditional practices of Livestock Keepers, and traditional breeds lose their specific characteristics once removed from these ecosystems.

> Livestock Keepers therefore have a right to access their natural environment, so as to ensure the sustainable use and conservation of their breeds and the environment.

**Principle 3:**
**Traditional breeds represent collective property, products of indigenous knowledge and cultural expression of Livestock Keepers.**

> While Livestock Keepers have collective custodianship rights over their breeds and the genetic traits of these breeds, it is crucial that these rights are supported and promoted at national levels. States must therefore respect, preserve and maintain the knowledge, innovations and practices of Livestock Keepers embodying lifestyles relevant for sustainable use and conservation of livestock diversity.

Based on these principles articulated and implicit in existing legal instruments and international agreements, Livestock Keepers from traditional livestock keeping communities and/or adhering to ecological principles of animal production, shall be given the following **Livestock Keepers' Rights**:

1. Livestock Keepers have the right to make breeding decisions and breed the breeds they maintain.
2. Livestock Keepers shall have the right to participate in policy formulation and implementation processes on animal genetic resources for food and agriculture.
3. Livestock Keepers shall have the right to appropriate training and capacity building and equal access to relevant services enabling and supporting them to raise livestock and to better process and market their products.
4. Livestock Keepers shall have the right to participate in the identification of research needs and research design with respect

to their genetic resources, as is mandated by the principle of Prior Informed Consent.

5. Livestock Keepers shall have the right to effectively access information on issues related to their local breeds and livestock diversity.

Such principles and rights must be entrenched in legally binding international and national instruments.

# Notes

## Epigraph

1. 'Herd', Wiktionary, 11 August, 2022, https://en.wiktionary.org/wiki/herd.
2. Charles Stépanoff et al., 'Animal Autonomy and Intermittent Coexistences: North Asian Modes of Herding', *Current Anthropology* 58, no. 1 (February 2017): 57–81 https://doi.org/10.1086/690120.

## Introduction

1. Michael Benanav, 'The Sheep Are Like Our Parents', *New York Times*, 27 July 2012, https://www.nytimes.com/2012/07/29/travel/following-a-navajo-sheep-herder.html.
2. Margarita Sachkova, 'COP26 or COPOUT26?PETA Marches for a Vegan World', PETA U.K., 6 November 2021, https://www.peta.org.uk/blog/cop26-march-vegan/.
3. Philippa Nuttal, 'Pat Brown: "Farm Animals Are the Most Destructive Technology on Earth"', *New Statesman*, updated 2 December 2021, https://www.newstatesman.com/the-environment-interview/2021/11/pat-brown-farm-animals-are-the-most-destructive-technology-on-earth.
4. For a map of pastoralists, see http://umap.openstreetmap.fr/de/map/pastoralists_563977#5/53.318/-7.053.
5. Jocelyne Porcher, 'The Work of Animals: A Challenge for Social Sciences', *Humanimalias* 6, no. 1 (2014): 1–9, https://doi.org/10.52537/humanimalia.9925.
6. Maryam Niamir-Fuller et al., 'Co-Existence of Wildlife and Pastoralism on Extensive Rangelands: Competition or Compatibility?', *Pastoralism* 2, no. 8 (2012): https://doi.org/10.1186/2041-7136-2-8.

## A Tapestry of Cultures

1. The Sheep from the Future, 'On the Spiritual Relationship between the Shepherd and His Herd', YouTube video, 3:00, 27 November 2021, https://www.youtube.com/watch?v=x-cVbG0GiSo.

2. Nomads, WAMIP, 'Uncle Sayyaad's Speech to the Plenary of the World Parks Congress', Cenesta, 9 September 2003, https://www.cenesta.org/en/2003/09/09/uncle-sayyaads-speech-to-the-plenary-of-the-world-parks-congress/: 'We, pastoral peoples, have always considered our land what you would call a "protected area". We have always embraced "conservation" not as a professional activity but as intimate duty and pride of every member of our tribes, as the heart of our livelihood, because our very subsistence depends on it. I hear you talk of ecosystems, landscapes and connectivity. We have always known about this without using your terms. Our migration patterns transfer seeds. Our grazing patterns shape the landscape. We subsist on our land; we know and care for its diversity of plants and animals. We pray on this land, and we guard its many sacred spaces. For the land provides us also with spiritual well-being.'
3. WAMIP, World Alliance of Mobile Indigenous Peoples, https://wamipglobal.com/.
4. 'Summary Report and Global Action Plan: Women Pastoralists', MARAG, November 2010, IUCN, https://iucn.org/sites/d/files/import/downloads/ggwp_summary_report_final_1.pdf.
5. European Commission Joint Research Centre, accessed on 14 December 2021, https://wad.jrc.ec.europa.eu/globalagriculture.
6. 'New Atlas Reveals Rangelands Cover Half the World's Land Surface, Yet Often Ignored Despite Threats', UNEP, 26 May 2021, https://www.unep.org/news-and-stories/press-release/new-atlas-reveals-rangelands-cover-half-worlds-land-surface-yet.
7. Nathan Sayre, *The Politics of Scale: A History of Rangeland Science* (Chicago: University of Chicago Press, 2017): 31.
8. Hyun Jin Kim, 'The Xiongnu', *Oxford Research Encyclopedias, Asian History,* last updated 29 March 2017, https://doi.org/10.1093/acrefore/9780190277727.013.50.
9. John Masson Smith, Jr, 'Review: Nomads on Ponies vs. Slaves on Horses', *Journal of the American Oriental Society* 118, no. 1 (1998): 54–62, https://doi.org/10.2307/606298.
10. See also: Max Liboiron, 'Decolonizing Geoscience Requires More than Equity and Inclusion', *Nature Geosciences* 14 (2 December, 2021): 876–877, https://doi.org/10.1038/s41561-021-00861-7.
11. Neeladri Bhattacharya, *The Great Agrarian Conquest: The Colonial Reshaping of a Rural World* (Delhi: Permanent Black, 2018).

12. Bhattacharya, *Great Agrarian Conquest.*
13. Sarah Cameron, 'The Kazakh Famine of 1930–33: Current Research and New Directions', *East/West Journal of Ukrainian Studies* 3, no. 2 (September 2016): 117, http://doi.org/10.21226/T2T59X.
14. Piers Vitebsky, 'Wild Tungus and the Spirits of Places', *Ab Imperio* (February 2012): 429–448, https://doi.org/10.1353/imp.2012.0046.
15. Mingming Fan et al., 'Impacts of Nomad Sedentarization on Social and Ecological Systems at Multiple Scales in Xinjiang Uyghur Autonomous Region, China', *Ambio* 43, no. 5 (2014): 673–686, http://doi.org/10.1007/s13280-013-0445-z.
16. 'Wasted Lives: A Critical Analysis of China's Campaign to End Tibetan Pastoral Lifeways', Tibetan Centre for Human Rights & Democracy, 2015, available at https://tchrd.org/wp-content/uploads/2022/07/Download-Report.pdf.
17. Joseph Poore and Thomas Nemecek, 'Reducing Food's Environmental Impacts Through Producers and Consumers', *Science* 360, no. 6392 (2018): 987–992, https://doi.org/10.1126/science.aaq0216.
18. Marie-Luise Hertkorn, Hassan Roba and Brigitte Kaufmann, 'Caring for Livestock: Borana Women's Perceptions of Their Changing Role in Livestock Management in Southern Ethiopia', *Nomadic Peoples* 19, no. 1 (2015): 30–52.
19. Piers Vitebsky, *Reindeer People: Living with Animals and Spirits in Siberia* (London: HarperCollins, 2005): 10–11.
20. Srijana Joshi et al., 'Ethnic and Cultural Diversity amongst Yak Herding Communities in the Asian Highlands', *Sustainability* 12 (2020): 957, https://doi.org/10.3390/su12030957.
21. Michael Benanav, *Himalaya Bound: One Family's Quest to Save Their Animals – And an Ancient Way of Life* (New York: Pegasus Books, 2018); see photo on VoA, 22 June 2018; images from Michael Benanav's journey with the Van Gujjars of Northern India, https://www.voanews.com/a/himalaya-bound/4450698.html.
22. Tegegne Teka. 'Camel and the Household Economy of the Afar'. *Nomadic Peoples* 29 (1991): 31–41. https://www.jstor.org/stable/43123336.
23. Pamela Burger, E. Ciani and B. Faye, 'Old World Camels in a Modern World – A Balancing Act Between Conservation and Genetic Improvement', *Animal Genetics* 50, no. 6 (2019): 598–612, https://doi.org/10.1111/age.12858.

24. Saverio Krätli, 'Cattle Breeding, Complexity and Mobility in a Structurally Unpredictable Environment: The WoDaaBe Herders of Niger', *Nomadic Peoples* 12, no. 1 (2008): 11–41.
25. Athanasios Ragko et al., 'Current Trends in the Transhumant Cattle Sector in Greece', *Scientific Papers: Animal Science and Biotechnologies* 46, no. 1 (2013): 422–426.
26. Ilse Köhler-Rollefson, 'The Shepherds that Worship Wolves', *Livestock Futures*, 23 February 2015, http://www.ilse-koehler-rollefson.com/?p=898.
27. Alessandro Puglia, 'Gli Asinelli Che Hanno Sconfitto la Mafia' [The Donkeys Who Defeated the Mafia], *Vita*, 21 February 2021, http://www.vita.it/it/story/2021/02/21/gli-asinelli-che-hanno-sconfitto-la-mafia/388/.
28. Grimaldo Vásquez, 'Culture and Biodiversity in the Andes', *Forest, Trees and People Newsletter* 34 (1997): 39–45.
29. Anita Sharma, *The Bakkarwals of Jammu and Kashmir: Navigating through Nomadism* (New-Delhi: Niyogi Books, 2009).
30. Oliver Milman, 'Five Hundred Goats Save the Ronald Reagan Library from Wildfires', *The Guardian*, 31 October 2019, https://www.theguardian.com/us-news/2019/oct/31/goats-save-ronald-reagan-library-wildfire.
31. 'Pastoralism', Your Dictionary, https://www.yourdictionary.com/pastoralism; 'Pastoralism', Vocabulary, https://www.vocabulary.com/dictionary/pastoralism.
32. Ruth Häckh, *Eine für Alle: Mein Leben als Schäferin* (München: Ludwig Buchverlag, 2018): 275. Translation by Ilse Köhler-Rollefson.
33. See Nikolaus Schareika, Christopher Brown and Mark Moritz, 'Critical Transitions from Pastoralism to Ranching in Central Africa', *Current Anthropology* 62, no. 1 (2021): 53–76 for an interesting discussion about the differences in relating to animals between herding and ranching.

## Knowledge of the Whole

1. SNS, 'Living Lightly Project Presents Online Exhibition 'Desi Oon' Focusing on Indian Indigenous Wool', The Statesman, 12 December 2020, https://www.thestatesman.com/features/living-lightly-project-presents-online-exhibition-desi-oon-focusing-indian-indigenous-wool-1502940397.html.
2. Maria Fernandez-Gimenez, 'The Role of Mongolian Nomadic Pastoralists; Ecological Knowledge in Rangeland Management', *Ecological Applications*

10, no. 5 (2000): 1318–1326, http://www.jstor.org/stable/2641287 for an overview of the knowledge system of Mongolian nomads.
3. For a detailed description of the plants that the camels eat, see LPPS, *The Camels of Kumbhalgarh: A Biodiversity Treasure* (Sadri, India: LPPS, 2013), http://www.lpps.org/wp-content/uploads/2013/10/Camels_Of_Kumbhalgarh_web.pdf.
4. H.E. Cross, *The Camel and its Diseases: Being Notes for Veterinary Surgeons and Commandants of Camel Corps* (London: Bailliere, Tindall & Cox, 1917).
5. Ilse Köhler-Rollefson, 'Between Burning Irons and Antibiotics', *Reports of the DFG*, no. 2–3 (1997): 4–6.
6. Zach Boren, 'Meat Industry Pushes UN Food Summit to Back Factory Farming', *Unearthed*, 29 September 2021, https://unearthed.greenpeace.org/2021/09/21/un-food-systems-summit-meat-climate/.

## Bonding

1. International Museum of the Horse, *A Gift from the Desert: The Art, History and Culture of the Arabian Horse* (2010): 67, https://archive.org/details/1GFDCatalogFront/page/n67/mode/2up?q=burckhardt.
2. Virginia DeJohn Anderson, *Creatures of Empire: How Domestic Animals Transformed Early America* (Oxford: Oxford University Press, 2004).
3. Summarised from Peter Iverson, *Diné: A History of the Navajos* (Albuquerque, NM: University of New Mexico Press, 2002) and Marsha Weisiger, *Dreaming of Sheep in Navajo Country* (Seattle: University of Washington Press, 2011).
4. Juliet Clutton-Brock, *Animals as Domesticates: A World View through History* (Ann Arbor, MI: University of Michigan Press, 2012).
5. Sandor Bökönyi, *The Domestication and Exploitation of Plants and Animals*, eds P.J. Ucko and G.W. Dimbleby (London: Duckworth, 1969): 219–229.
6. Hans-Peter Uerpmann, 'Animal Domestication ± Accident or Intention?', *The Origin and Spread of Agriculture and Pastoralism in Eurasia*, ed. D.R. Harris (London: University College London Press, 1996): 227–237.
7. Piers Vitebsky, *Reindeer People: Living with Animals and Spirits in Siberia* (London: HarperCollins, 2005): 10–11.
8. Saverio Krätli, 'Cows Who Choose Domestication: Generation and Management of Domestic Animal Diversity by WoDaaBe Pastoralists (Niger)', PhD thesis, Institute of Development Studies at the University of Sussex U.K., 2007.

9. More information about 'Ain Ghazal in Alan Simmons et al., 'Ain Ghazal: A Major Neolithic Settlement in Central Jordan', *Science* 240, 4848 (1988): 35–39, https://www.science.org/doi/10.1126/science.240.4848.35.
10. Ilse Köhler-Rollefson, 'Changes in Goat Exploitation at 'Ain Ghazal between the Early and Late Neolithic: A Metrical Analysis', *Paléorient* 15, no. 1 (1989): 141–146, https://doi.org/10.3406/paleo.1989.4492.
11. This was in the context of an ethnoarchaeological project to investigate how nomadic occupation is reflected in the archaeological record. See Edward Banning and Ilse Köhler-Rollefson, 'Ethnoarchaeological Survey in the Beidha Area, Southern Jordan', *Annual of the Department of Antiquity of Jordan* 27 (1983): 375–384.
12. Ilse Köhler-Rollefson and Gary Rollefson, 'Brooding about Breeding: Social Implications for the Process of Animal Domestication, *The Dawn of Farming in the Near East*, eds R.T.J. Cappers and S. Bottema (Berlin: Ex Oriente, 2002): 177–182.
13. Natasha Fijn, *Living with Herds: Human-Animal Co-Existence in Mongolia* (Cambridge: Cambridge: University Press, 2011).
14. Charles Stépanoff et al., 'Animal Autonomy and Intermittent Co-existences: North Asian Modes of Herding', *Current Anthropology* 58, no. 1 (February 2017): 57–81.
15. Penny Dransart, *Earth, Water, Fleece and Fabric: An Ethnography and Archaeology of Andean Camelid Herding* (London: Routledge, 2002).
16. Alex Blanchette, *Porkopolis: American Animality, Standardized Life, and the Factory Farm* (Durham: Duke University Press, 2020).
17. Gary Rollefson, personal communication, 25 December 2022.
18. Gary Rollefson, personal communication, 22 December 2021.

## Communication

1. Madhuram Raika, personal communication, 10 October 2021.
2. Saverio Krätli, 'Cattle Breeding, Complexity and Mobility in a Structurally Unpredictable Environment: The WoDaaBe Herders of Niger', *Nomadic Peoples* 12, no. 1 (2008): 11–41.
3. Kip Hutchins, 'Like a Lullaby: Song as Herding Tool in Rural Mongolia', *Journal of Ethnobiology* 39, no. 3 (2019): 445–459, https://doi.org/10.2993/0278-0771-39.3.445.
4. Anna Ivarsdotter, 'And the Cattle Follow Her, for They Know Her Voice... of Communication between Women and Cattle in Scandinavian

Pastures', in *PECUS Man and Animals in Antiquity: Proceedings of the Conference at the Swedish Institute in Rome*, 9–12 September 2002, ed. Barbro Santillo Frizell (Rome: The Swedish Institute in Rome, 2004): 146–149 and Camilla Eriksson, 'What Is Traditional Pastoral Farming? The Politics of Heritage and "Real Values" in Swedish Summer Farms (*Fäbodbruk*)', *Pastoralism* 125 (2011), https://doi.org/10.1186/2041-7136-1-25.

5. Paul Riesman, *Freedom in Fulani Social Life: An Introspective Ethnography* (Chicago: University of Chicago Press, 1977): 102.
6. Florian Stammler, 'Animal Diversity and its Social Significance among Arctic Pastoralists', *Good to Eat, Good to Live with: Nomads and Animals in Northern Eurasia and Africa*, eds Florian Stammler and Hiroki Takakura (Sendai, Japan: Center for Northeast Asian Studies, Tohoku University, 2010): 215–243.
7. Mark the Jura, '*Oameni şi câini. O poveste adevărată despre nişte ciobăneşti legendari'* [People and Dogs: A True Tale about Legendary Sheep Dogs], *Marcu Jura Republica*, 2 January 2016, http://republica.ro/oameni-c-i-caini-o-poveste-adevarata-despre-nic-te-ciobanec-tilegendari.
8. Zoom interview with Sylvain Perdigon, 15 November 2021.

## Transformation or the Difference Between Herding and Farming

1. Arte Film on Herders, https://en.neue-celluloid-fabrik.de/filme/herders-wt/.
2. Ilse Köhler-Rollefson and Hanwant Singh Rathore, 'The Case of the Kumbhalgarh Wildlife Sanctuary and Camel Pastoralism in Rajasthan (India)', *Sustainability* 13, no. 24 (2021): 13914, https://doi.org/10.3390/su132413914.
3. See opinion piece by Atul Chaturvedi, 'Improving Livestock Breeding', *The Hindu*, 19 October 2021, https://www.thehindu.com/opinion/op-ed/improving-livestock-breeding/article37060492.ece.
4. More information about the Pathe Pathshala and a link to the video about the Chilika buffalo at http://pathepathshala.org/video.html.
5. More information: Dr Balaram Sahu, 'Pig Pastoralism in Odisha – A Study', http://pathepathshala.org/books/Pig%20Pastoralism%20in%20Odisha%20F-3.pdf.
6. Living Lightly, http://livinglightly.in/.
7. R.B. Ekvall, *Fields on the Hoof: Nexus of Tibetan Nomadic Pastoralism* (New York: Holt, Rinehart and Winston, 1968).

8. Be Obua, 'Checklist of Forage Plants Utilized for Sheep Feeding in Nsukka Area of Enugu State, Southeastern Nigeria', *National Journal of Agriculture and Rural Development* 21, no.2 (2018): 3548–3565.
9. Be Obua, 'Survey of the Diversity of Browse Plants Utilized for Goat Feeding in Ohaji/Egbema/Oguta Area of Imo State, Nigeria', *International Journal of Tropical Agriculture and Food Systems* 7, no. 1 (2013): 54–66.
10. Y. Geng et al., 'Prioritizing Fodder Species Based on Traditional Knowledge: A Case Study of Mithun (Bos frontalis) in Dulongjiang area, Yunnan Province, Southwest China', *Journal of Ethnobiology and Ethnomedicine* 13, no. 24 (2017): https://doi.org/10.1186/s13002-017-0153-z.
11. Georg Klute, *Die Schwerste Arbeit der Welt. Alltag von Tuareg Nomaden* (Wuppertal, Germany: Trickster Verlag, 1992); Shun Sato, 'Pastoral Movements and the Subsistence Unit of the Rendille of Northern Kenya: with Special Reference to Camel Ecology', *Senri Ethnological Studies* 6 (1980): 1–178.
12. Food and Agriculture Organization of the United Nations, *The State of the World's Biodiversity for Food and Agriculture*, eds Julie Bélanger and Dafydd Pilling (FAO Commission on Genetic Resources for Food and Agriculture Assessments: Rome, 2019), http://www.fao.org/3/CA3129EN/CA3129EN.pdf.
13. Food and Agriculture Organization of the United Nations. *Pastoralism – Making Variability Work*. FAO Animal Production and Health Paper No. 185. Rome: FAO, 2021. https://doi.org/10.4060/cb5855en.

## Movement

1. 'The Indigenous Cattle Breeds of Rajputana' (Calcutta, Government Printing, 1909).
2. Brett Jesmer et al., 'Is Ungulate Migration Culturally Transmitted? Evidence of Social Learning from Translocated Animals', *Science* 361, no. 6406 (2018): 1023–1025, https://dx.doi.org/10.1126/science.aat0985.
3. Marco Festa Bianchet, 'Learning to Migrate', *Science* 361 (2018): 972–997, https://doi.org/10.1126/science.aau6835.
4. Christer M. Rolandsen et al., 'On Fitness and Partial Migration in a Large Herbivore – Migratory Moose Have Higher Reproductive Performance Than Residents', *Oikos* 126, no. 4 (2017): 547–555.
5. *Cañadas*: The name *cañadas reales* refers to those paths traditionally used for transhumance in Spain, regulated by royal edict of Alfonso X the Wise in 1273. Although the trails laid out by the *cañadas*, later known as

*cañadas reales*, were routes used from ancient times by transhumance shepherds, Alfonso X's decree sought to regulate, organise and protect certain trails that, due to their importance, use or location, deserved to be preserved from possible violations. Thus, together with the creation of the *Concejo de la Mesta*, the *cañadas reales* were defined. See also Lisa Zogib, 'On the Move for 10000 Years: Biodiversity Conservation through Transhumance and Nomadic Pastoralism in the Mediterranean', Alliance for Mediterranean Nature and Culture, 2014, http://medconsortium.org/wpcontent/uploads/2018/01/10000Years_MediterraneanConsortiumForNatureAndCulture.pdf.

6. Saverio Krätli, 'Cows Who Choose Domestication: Generation and Management of Domestic Animal Diversity by WoDaaBe Pastoralists (Niger)', PhD dissertation, Institute of Development Studies at the University of Sussex, U.K., 2007.
7. Saverio Krätli, Christian Huelsebusch, Sally Brooks, Brigitte Kaufmann, 'Pastoralism: A Critical Asset for Food Security under Global Climate Change', *Animal Frontiers* 3, no. 1 (January 2013): 42–50, https://do.org/10.2527/af.2013-0007.
8. Georg Klute, *Die Schwerste Arbeit der Welt. Alltag von Tuareg Nomaden* (Wuppertal, Germany: Trickster Verlag, 1992).

## Nourishment

1. Gene Logsdon, *Holy Shit: Managing Manure to Save Mankind* (White River Junction: Chelsea Green, 2010).
2. Jared Diamond, *Collapse: How Societies Choose to Fail or Succeed* (London: Penguin Books, 2011).
3. Gene Logsdon, *Holy Shit*.
4. 'Mama' is an honorific title, meaning maternal uncle. This is the way he is referred to by anybody who knows him, so I am also adopting this practice.
5. J. Kotschi, *A Soiled Reputation: Adverse Impacts of Mineral Fertilizers in Tropical Agriculture* (Berlin: Heinrich Böll Foundation and WWF, Germany, 2013).
6. Vandana Shiva, 'In Praise of Cowdung', *Znet*, 20 November 2002, https://zcomm.org/znetarticle/in-praise-of-cowdung-by-vandana2-shiva-1/.
7. Sir Albert Howard, 'The Animal as Our Farming Partner', *Organic Gardening* 2, no. 3, (September 1947): http://journeytoforever.org/farm_library/howard_animal.html.

8. B. Sriveda and B. Srihitha, 'Sheep Penning: Need to Sustain this Unique Practice', Leisa India, accessed 30 May 2022, https://leisaindia.org/sheep-penning-need-to-sustain-this-unique-practice.
9. Christopher Delgado, Mark Rosegrant and Henning Steinfeld, 'Livestock to 2020: The Next Food Revolution' *Outlook on Agriculture* 30, no.1 (2001): 27–29, https://doi.org//10.5367/000000001101293427.
10. Joe Wertz, 'How Big Farms Got a Big Government Pass on Air Pollution', Center for Public Integrity, 16 September 2020, https://publicintegrity.org/environment/factory-farming-air-pollution-pass-cafos/.
11. Rebecca Hersher, 'Hope and Skepticism as Biden Promises to Address Environmental Racism', NPR,29 January 2021, https://www.npr.org/2021/01/29/956012329/hope-and-skepticism-as-biden-promises-to-address-environmental-racism.
12. L. Hekman et al., '"Toilet of Europe": Spain's Pig Farms Blamed for Mass Fish Die-Off', *The Guardian*, 13 October 2021, https://www.theguardian.com/environment/2021/oct/13/toilet-of-europe-spains-pig-farms-blamed-for-mass-fish-die-offs.
13. Sam Jones, 'Spanish Should Eat Less Meat to Limit Climate Crisis, Says Minister', *The Guardian*, 26 December 2021, https://www.theguardian.com/world/2021/dec/26/spanish-should-eat-less-meat-to-limit-climate-crisis-says-minister.
14. 'How a Minister's "Poor Meat" Comments in "The Guardian" Triggered a Political Storm in Spain', *El Pais*, 12 January 2022, https://english.elpais.com/spain/2022-01-12/how-a-ministers-poor-meat-comments-in-the-guardian-triggered-a-political-storm-in-spain.html.
15. 'International Coalition Petitions Inter-American Commission on Human Rights to Investigate Factory-Farm Abuses', Waterkeeper Alliance, 11 October 2021, https://waterkeeper.org/news/international-coalition-petitions-inter-american-commission-on-human-rights-to-investigate-factory-farm-abuses/.

## Creating Breed Diversity

1. Dessie and Mwai, eds, *The Story of Cattle in Africa: Why Diversity Matters* (Nairobi, Kenya: International Livestock Research Institute, 2019).
2. Kor Oldenbroek and Liesbeth van der Waaij, 'Textbook Animal Breeding: Animal Breeding and Genetics for BSc Students', 2014, Wageningen University and Research Centre, the Netherlands, https://www.wur.nl/

upload_mm/4/d/5/3b256d1f-2ae5-4fa6-a2d7-afe355ce9870_Textbook%20Animal%20Breeding%20and%20Genetics-v17-20160105_2020.pdf.

3. Julius Klein, *The Mesta: A Study in Spanish Economic History* 1273–1836 (Harvard University Press, 1920): 4–5.
4. International Museum of the Horse, *A Gift from the Desert: The Art, History and Culture of the Arabian Horse*, 2010, https://archive.org/details/1GFDCatalogFront/page/n17/mode/2up.
5. Alois Musil, *The Manners and Customs of the Rwala Bedouins* (New York: The American Geographical Society, 1928).
6. John Galaty, 'Cattle and Cognition: Aspects of Maasai Practical Reasoning', in *The Walking Larder: Patterns of Domestication, Pastoralism, and Predation*, ed. J. Clutton-Brock (London: Unwin Hyman, 1989).
7. Christian Hülsebusch et al., *Camel Breeds and Breeding in Northern Kenya* (Nairobi, Kenya: Kenya Agricultural Research Institute, 2002).
8. A. Elmi, *Camel Husbandry and Management by Celdheer Pastoralists in Central Somalia*, Pastoral Development Network Paper 27d (London: Overseas Development Institute, 1989).
9. Brian Hartley, 'The Dromedary of the Horn of Africa', *The Camelid: An All-Purpose Animal, Vol. I.*, ed. W.R. Cockrill (Uppsala: Scandinavian Institute of African Studies, 1984), 77–97.
10. A. Elmi, *Camel Husbandry*.
11. I.M. Lewis, *A Pastoral Democracy: A Study of Pastoralism and Politics among the Northern Somali of the Horn of Africa* (Oxford: Oxford University Press, 1961): 85.
12. Ilse Köhler-Rollefson, 'Indigenous Practices of Animal Genetic Resource Management and Their Relevance for the Conservation of Domestic Animal Diversity in Developing Countries', *Journal of Animal Breeding and Genetics* 114 (1997): 231–238, https://pubmed.ncbi.nlm.nih.gov/21395819/.
13. Ellen Geerlings, 'The Black Sheep of Rajasthan', *Seedling* (2004): https://grain.org/article/entries/436-the-black-sheep-of-rajasthan.
14. Günther Schlee, 'Camel Management Strategies and Attitudes towards Camels in the Horn', *The Exploitation of Animals in Africa*, ed. J. Stone (Aberdeen: Aberdeen University, African Studies Group, 1989): 143–154.
15. For instance, see the standard for Herdwick sheep at https://web.archive.org/web/20120204073104/http://www.herdwick-sheep.com/herdwick_breed_standard.html.

16. Breeds of Livestock, Department of Animal Science, Oklahoma State University, accessed on 30 May 2022, http://afs.okstate.edu/breeds/.
17. Jay Lush, 'The Genetics of Populations', quoted on Breeds of Livestock website http://afs.okstate.edu/breeds/.
18. Ilse Köhler-Rollefson, 'Indigenous Practices'.

## Stewarding Biological Diversity

1. Jesús Garzón-Heydt, personal communication (September 2007, at foundation meeting of WAMIP in Segovia, Spain).
2. Anil Chhangani, Paul Robbins, and S.M. Mohnot, 'Crop Raiding and Livestock Predation at Kumbhalgarh Wildlife Sanctuary, Rajasthan, India', *Human Dimensions of Wildlife* 13 (2008): 305-316. https://doi.org/10.1080/10871200802282922.
3. Mohit M. Rao, 'Why There's no Conflict between Wolves and Shepherds in These Koppal Villages', *The Hindu*, 22 December 2018, https://www.thehindu.com/sci-tech/energy-and-environment/why-theres-no-conflict-between-wolves-and-shepherds-in-these-koppal-villages/article25789757.ece.
4. Ilse Köhler-Rollefson, 'The Shepherds That Worship Wolves', *Livestock Futures,* 23 February 2015, http://www.ilse-koehler-rollefson.com/?p=898. See also Nitya Ghotge and Sagari Ramdas, 'Black Sheep and Gray Wolves': http://www.anthra.org/wp-content/uploads/2016/11/Black-Sheep-and-Gray-Wolves.pdf in Seminar.
5. Iravatee Majgaonkar et al., 'Land-Sharing Potential of Large Carnivores in Human-Modified Landscapes of Western India', *Conservation Science and Practice* 1, no. 5 (2019): e34, https://doi.org/10.1111/csp2.34.
6. 'Put to the Horn', Down to Earth, 15 July 1998, https://www.downtoearth.org.in/blog/put-to-the-horn-22005.
7. Michael Lewis, 'Cattle and Conservation at Bharatpur: A Case Study in Science and Advocacy', *Conservation and Society* 1, no. 1 (2003): 1–21, http://www.jstor.org/stable/26396448.
8. Steve Cracknell, *The Implausible Rewilding of the Pyrenees* (Lulu.com, 2021).
9. Pedro Oleaa and Patricia Mateo-Tomás, 'The Role of Traditional Farming Practices in Ecosystem Conservation: The Case of Transhumance and Vultures', *Biological Conservation* 142 (2009): 1844–1853 , https://doi.org/10.1016/j.biocon.2009.03.024.

10. *Trashumancia y Naturaleza*, https://trashumanciaynaturaleza.org/en-gb/home.
11. Dr Paul Starrs, 'Transhumance as Antidote for Modern Sedentary Stock Raising', *Rangeland Ecology & Management* 71, no. 5 (2018): 592–602, https://doi.org/10.1016/j.rama.2018.04.011.
12. Pablo Manzano and J. Malo, 'Extreme Long-Distance Seed Dispersal Via Sheep', *Frontiers in Ecology and the Environment* 4, no. 5 (2006): 244–248, https://www.jstor.org/stable/3868790.
13. S. Fischer, P. Poschlod and B. Beinlich, 'Experimental Studies on the Dispersal of Plants and Animals on Sheep in Calcareous Grasslands', *Journal of Applied Ecology* 33, no. 5 (2006): 1206–1222, https://doi.org/10.2307/2404699.
14. A. Koocheki, 'Herders Care for Their Land', *ILEIA Newsletter* 8, no. 3 (1992): 3.
15. Herbert Nickel, 'Zikaden', in M. Bunzel-Drüke et al., *Naturnahe Beweidung und NATURA 2000* (Bad Sassendorf, Germany: Arbeitsgemeinschaft Biologischer Umweltschutz, 2019): 267–277.
16. Chen Dongming et al., 'The Effect of Different Restoration Measures on the Desertified Alpine Grassland in Zoigê', *Chinese Journal of Applied & Environmental Biology* 22 (2016): 573–578.
17. Peter Poschlod, 'The Historical and Socioeconomic Perspective of Calcareous Grasslands – Lessons from the Distant and Recent Past', *Biological Conservation* 104, no. 3 (2002): 361–376, https://doi.org/10.1016/S0006-3207(01)00201-4.
18. Jan Butaye, Dries Adriaens and Olivier Honnay, 'Conservation and Restoration of Calcareous Grasslands: A Concise Review of the Effects of Fragmentation and Management on Plant Species', *Biotechnology, Agronomy, Society, Environment* 9, no. 2 (2005): 111–118.
19. European Commission, 'Improving the Management of Salisbury Plain Natura 2000 Sites', accessed 24 April 2022, https://webgate.ec.europa.eu/life/publicWebsite/index.cfm?fuseaction=search.dspPage&n_proj_id=1712.
20. Fiona Marshall et al., 'Ancient Herders Enriched and Restructured African Grasslands', *Nature* 561, no. 7723 (2018): 387–390, https://doi.org/10.1038/s41586-018-0456-9.
21. Michael Spate et al., 'Palaeoenvironmental Proxies Indicate Long-Term Development of Agro-Pastoralist Landscapes in Inner Asian

Mountains', *Scientific Reports* 12 (2022): 554, https://www.nature.com/articles/s41598-021-04546-4.

22. Alicia Ventresca Miller et al., 'Ecosystem Engineering Among Ancient Pastoralists in Northern Central Asia', *Frontiers Earth Science* 8 (2020): 168, https://doi.org/10.3389/feart.2020.00168.

## Regenerating Land

1. Albert Howard, *An Agricultural Testament* (New York and London: Oxford University Press, 1943).
2. Allan Savory, 'How to Green the World's Deserts and Reverse Climate Change', YouTube video, 22:19, https://www.youtube.com/watch?v=vpTHi7O66pI.
3. Steven Apfelbaum et al., 'Vegetation, Water Infiltration, and Soil Carbon Response to Adaptive Multi-Paddock and Conventional Grazing in Southeastern USA Ranches', *Journal of Environmental Management* 308 (April 2022): 114576, https://doi.org/10.1016/j.jenvman.2022.114576.
4. Roy Behnke and Michael Mortimore, eds, *The End of Desertification? Disputing Environmental Change in the Drylands* (Berlin: Springer, 2016).
5. UNEP, 'Back to the Future: Rangeland Management in Jordan', *UNEP News and Stories*, 22 June 2016, https://www.unenvironment.org/news-and-stories/story/back-future-rangeland-management-jordan.
6. Elinor Ostrom et al., eds, *The Drama of the Commons* (Washington, D.C.: National Academies Press, 2002).
7. M.A. Seid et al., 'The Role of Pastoralism in Regulating Ecosystem Services', *Revue Scientifique et Technique (International Office of Epizootics)* 35, no. 2 (2016): 435–444, https://doi.org/10.20506/rst.35.2.2534.
8. Hilde Gauthier-Pilters and Anne I. Dagg, *The Camel: Its Evolution, Ecology, Behaviour and Relationship to Man* (Chicago: Chicago University Press, 1981).
9. Kyinzom Dhongdue, Gabriel Lafitte and Simon Bradshaw, *An Iron Fist in a Green Glove: Emptying Pastoral Tibet with China's National Parks* (Australia Tibet Council, 2019), https://www.atc.org.au/wp-content/uploads/2019/06/An-Iron-Fist-in-a-Green-Glove_online.pdf.
10. Gerald Roche, 'Abandoning the High Ground: The Ecological Implications of Pastoral Abandonment in Tibet', presented at the 3rd Himalayan Studies Conference, Yale (14–16 March 2014); Dhongdue et al., *Iron Fist*.

11. Mesfin Mekonnen and Arjen Hoekstra, 'A Global Assessment of the Water Footprint of Farm Animals and Animal Products', *Ecosystems* 15, no. 3 (2012): 401–415, https://doi.org/10.1007/s10021-011-9517-8.
12. Angeles Crest Creamery, https://www.angelescrestcreamery.com/.

## Cooling the Climate

1. R. K. Heitschmidt et al., 'Ecosystems, Sustainability, and Animal Agriculture', *Journal of Animal Science*, 74, no. 6 (June 1996): 1395–1405, https://doi.org/10.2527/1996.7461395x.
2. Food and Agriculture Organization of the United Nations, *Livestock's Long Shadow: Environmental Issues and Options* (Rome: FAO, 2006).
3. United Nations, 'Rearing Cattle Produces More Greenhouse Gases than Driving Cars, UN Report Warns', UN News, 26 November 2006, https://news.un.org/en/story/2006/11/201222-rearing-cattle-produces-more-greenhouse-gases-driving-cars-un-report-warns.
4. Henning Steinfeld, Pierre Gerber and Carolyn Opio, 'Responses on Environmental Issues', *Livestock in a Changing Landscape: Drivers, Consequences, and Responses*, eds Henning Steinfeld et al. (Washington, D.C.: Island Press, 2010): 313.
5. Petr Havlík et al., 'Climate Change Mitigation through Livestock System Transitions', *Proceedings of the National Academy of Sciences* 111, no. 10 (2014): 3709–3714, https://doi.org/10.1073/pnas.1308044111.
6. Rod Heitschmidt, R.E. Short and E.E. Grings, 'Ecosystems, Sustainability, and Animal Agriculture', *Journal of Animal Science* 74, no. 6 (1996): 1395–1405.
7. W.R. Teague et al., 'The Role of Ruminants in Reducing Agriculture's Carbon Footprint in North America', *Journal of Soil and Water Conservation* 71, no. 2 (2016): 156–164, https://doi.org/10.2489/jswc.71.2.156.
8. Albert Howard, *An Agricultural Testament* (New York and London: Oxford University Press, 1943).
9. Pierre Gerber et al., *Tackling Climate Change through Livestock – A Global Assessment of Emissions and Mitigation Opportunities*, Rome: Food and Agriculture Organization of the United Nations, 2013.
10. Tara Garnett et al., *Grazed and Confused? Ruminating On Cattle, Grazing Systems, Methane, Nitrous Oxide, the Soil Carbon Sequestration Question – And What It All Means for Greenhouse Gas Emissions* (Oxford: FCRN, University of Oxford, 2017), https://www.oxfordmartin.ox.ac.uk/downloads/reports/fcrn_gnc_report.pdf.

11. Rishika Pardikar, 'Large Herbivores May Improve an Ecosystem's Carbon Persistence', *Eos* 103 (2022), https://doi.org/10.1029/2022EO220029.
12. Ella Houzer and Ian Scoones, *Are Livestock Always Bad for the Planet? Rethinking the Protein Transition and Climate Change Debate* (Brighton: Institute of Development Studies, 2021), https://pastres.org/livestock-report/.
13. S. Park et al., 'Trends and Seasonal Cycles in the Isotopic Composition of Nitrous Oxide since 1940', *Nature Geoscience* 5 (2012): 261–265, https://doi.org/10.1038/ngeo1421.
14. Myles Allen et al., 'A Solution to the Misrepresentations of $CO_2$-Equivalent Emissions of Short-Lived Climate Pollutants under Ambitious Mitigation', *NPJ Climate and Atmospheric Science* 1, no. 16 (2018): https://doi.org/10.1038/s41612-018-0026-8.
15. Houzer and Scoones, *Are Livestock Always Bad for the Planet?*
16. Michelle Cain et al., 'Improved Calculation of Warming-Equivalent Emissions for Short-Lived Climate Pollutants', *NPJ Climate and Atmospheric Science* 2, no. 29 (2019): https://doi.org/10.1038/s41612-019-0086-4.
17. Anthony N. Hristov, 'Historic, Pre-European and Present-Day Contribution of Wild Ruminants to Enteric Methane Emissions in the United States', *Journal of Animal Science* 90 (2012): 1371–1375, https://pubmed.ncbi.nlm.nih.gov/22178852.
18. G. Czerkus, personal communication, email on 24 November 2021.
19. Coral Murphy Marcos, 'The Unconventional Weapon Against Future Wildfires: Goats', *New York Times*, 18 September 2021, https://www.nytimes.com/2021/09/18/business/wildfires-goats-prevention.html.
20. 'Reindeer as Ecosystem Engineers?' 8 July 2021, Earth Observatory, https://earthobservatory.nasa.gov/images/149246/reindeer-as-ecosystem-engineers?fbclid=IwAR1X1kYptwGjqOUCzA2WNnqRBq4u1ANM5Ib-94JJJZdncnhpcjipQL1fcDI.
21. Megha Verma et al., 'Can Reindeer Husbandry Management Slow Down the Shrubification of the Arctic?' *Journal of Environmental Management* 267 (2020): 110636, https://doi.org/10.1016/j.jenvman.2020.110636.


## Feeding the World


1. Hilde Gauthier-Pilters and Anne Dagg, *The Camel: Its Evolution, Ecology, Behaviour and Relationship to Man* (Chicago: Chicago University Press, 1981).

2. Elizabeth Buff, 'Can We Solve World Hunger and Feed 9 Billion People Just By Eating Less Meat?', One Green Planet, https://www.onegreenplanet.org/environment/world-hunger-population-growth-ditching-meat/.
3. Joseph Poore and Thomas Nemecek, 'Reducing Food's Environmental Impacts Through Producers and Consumers', *Science* 360 (2018): 987–992, https://doi.org/10.1126/science.aaq0216.
4. Food and Agriculture Organization of the United Nations, *Pastoralism in Africa's Drylands* (Rome: FAO, 2018).
5. Adegbola Adesogan et al., 'Animal Source Foods: Sustainability Problem or Malnutrition and Sustainability Solution? Perspective Matters', *Global Food Security* 25 (2020): 100325, https://doi.org/10.1016/j.gfs.2019.100325.
6. D.T. Thomas et al., 'Net Protein Contribution and Enteric Methane Production of Pasture and Grain-Finished Beef Cattle Supply Chains', *Animal* 5, no. 12 (2021): 100392, https://doi.org/10.1016/j.animal.2021.100392.
7. Grassland 2.0, https://grasslandag.org.
8. Hannah Van Zanten et al., 'The Role of Farm Animals in a Circular Food System', *Global Food Security* 21 (2019): 18–22, https://doi.org/10.1016/j.gfs.2019.06.003.
9. Hannah Van Zanten et al., 'Defining a Land Boundary for Sustainable Livestock Consumption', *Global Change Biology* 9 (2018): 4185–4194, https://doi.org10.1111/gcb.14321.
10. CLEAR Center (Clarity and Leadership for Environmental Awareness and Research at UC Davis), 'Dairy Cows – The Original Upcyclers: How Ruminant digestion turns byproducts into high-quality nutrition', 7 January 2022, https://clear.ucdavis.edu/explainers/dairy-cows-original-upcyclers.
11. African Union Department of Rural Economy and Agriculture, *Policy Framework for Pastoralism in Africa: Securing, Protecting and Improving the Lives, Livelihoods and Rights of Pastoralist Communities* (Addis Ababa: African Union, 2010), https://au.int/sites/default/files/documents/30240-doc-policy_framework_for_pastoralism.pdf.

## Balancing Nutrition

1. Fred Provenza, Michel Meuret and Pablo Gregorini, 'Our Landscapes, Our Livestock, Ourselves: Restoring Broken Linkages among Plants, Herbivores, and Humans with Diets that Nourish and Satiate', *Appetite*, 2015: 513.

2. R.P. Agrawal et al., 'Zero Prevalence of Diabetes in Camel Milk Consuming Raica Community of North-West Rajasthan, India', *Diabetes Research and Clinical Practice* 76, no. 2 (2007): 290–296, https://doi.org/10.1016/j.diabres.2006.09.036.
3. Kush Purohit, Hanwant Singh Rathore and Ilse Köhler-Rollefson, 'Increased Risk of Type 2 Diabetes Mellitus in the Maru Raika Community of Rajasthan: A Cross-Sectional Study', *International Journal of Diabetes in Developing Countries* 37 (2016): 494–450, http://dx.doi.org/10.1007/s13410-016-0529-y.
4. Christina Adams, *Camel Crazy: A Quest for Miracles in the Mysterious World of Camels* (Novato, CA: New World Library, 2019).
5. Provenza et al., 'Our Landscapes'.
6. 'Fue aprobada a nivel nacional la denominacion de origen del Chivito Criollo del Norte Neuquino' [The Denomination of Origin of the Chivito Criollo de Norte Neuquino Was Approved at the National Level], *Senado Argentina*, June 24, 2010, https://www.senado.gob.ar/prensa/8606/noticias

## Death

1. Häckh, Ruth, personal communication via email, 15 July 2022. Translation by Ilse Köhler-Rollefson.
2. Marianne Landzettel, 'The Kindness of Butchers – Why a Small, Farmer-Owned Abattoir in Southern Germany Is Thriving', Sustainable Food Trust, 3 December 2021, https://sustainablefoodtrust.org/news-views/the-kindness-of-butchers-pt-1/.

## Conclusion

1. Alice Walker; 'Foreword', *The Dreaded Comparison: Human and Animal Slavery* by Marjorie Spiegel, Mirror Books, 1989.
2. John Vidal, 'Health Risks of Shipping Pollution Have Been Underestimated', *The Guardian*, 9 April 2009, https://www.theguardian.com/environment/2009/apr/09/shipping-pollution.
3. For a case-study, see Ilse Köhler-Rollefson and Hanwant Singh Rathore, 'The Case of the Kumbhalgarh Wildlife Sanctuary and Camel Pastoralism in Rajasthan (India)', *Sustainability* 13, no. 24 (2021): 13914, https://doi.org/10.3390/su132413914.
4. Emery Roe, 'A New Policy Narrative for Pastoralism? Pastoralists as Reliability Professionals and Pastoralist Systems as Infrastructure',

STEPS Working Paper No. 113. (Brighton, UK: Social, Technological and Environmental Pathways to Sustainability (STEPS) Centre, Institute of Development Studies, 2020).

5. Food and Agriculture Organization of the United Nations, 'Pastoralism – Making Variability Work,' FAO Animal Production and Health Paper No. 185 (Rome: FAO, 2021): https://doi.org/10.4060/cb5855en.
6. 'Managing the Complexities of Land & Livestock', Savory Global, last accessed 21 September, 2022, https://savory.global/holistic-management/.
7. D.M. Nyariki and D.A. Amwata, 'The Value of Pastoralism in Kenya: Application of Total Economic Value Approach', *Pastoralism* 9, no. 9 (2019): https://doi.org/10.1186/s13570-019-0144-x.
8. Helen de Jode, *The Green Quarter: A Decade of Progress Across the World in Sustainable Pastoralism* (Nairobi: IUCN, 2014), https://www.iucn.org/resources/publication/green-quarter-decade-progress-across-world-sustainable-pastoralism.
9. Marco Bassi, 'Pastoralists are Peoples: Key Issues in Advocacy and the Emergence of Pastoralists' Rights', *Nomadic Peoples* 21, no. 1 (2017): 4–33.
10. 'The Karen Commitment: Proceedings of a Conference of Indigenous Livestock Breeding Communities on Animal Genetic Resources', League for Pastoral Peoples, http://www.pastoralpeoples.org/wp-content/uploads/2020/01/karen1.pdf.
11. Les Bergers Urbains, https://www.bergersurbains.com/.

## Appendix 1

1. Nomads, WAMIP, 'Uncle Sayyaad's Speech to the Plenary of the World Parks Congress', 9 September 2003, Cenesta, http://www.cenesta.org/en/2003/09/09/uncle-sayyaads-speech-to-the-plenary-of-the-world-parks-congress/.
2. I.U. Köhler-Rollefson et al., 'Livestock Keepers' Rights: The State of Discussion', *Animal Genetic Resources/Resources génétiques animales/Recursos genéticos animales*, vol. 47 (July 2010): 119–123, https://doi.org/10.1017/S2078633610000925.
3. 'International Treaty on Plant Genetic Resources, Farmers' Rights', Food and Agriculture Organization of the United Nations, https://www.fao.org/plant-treaty/areas-of-work/farmers-rights/en/.
4. LPP, 'The Karen Commitment: Pastoralist/Indigenous Livestock Keepers' Rights', League for Pastoral Peoples and Endogenous Livestock

Development, http://www.pastoralpeoples.org/documents/the-karen-commitment-pastoralist-indigenous-livestock-keepers-rights/.

5. 'Wilderswil Declaration on Livestock Diversity', *La Via Campesina*, https://viacampesina.org/en/wilderswil-declaration-on-livestock-diversity/.
6. Food and Agriculture Organization of the United Nations. *Global Plan of Action for Animal Genetic Resources and the Interlaken Declaration* (Rome: FAO, 2007). https://www.fao.org/3/a1404e/a1404e.pdf.
7. LPP, 2018, *Community protocols for pastoralists and livestock keepers: Claiming rights under the Convention on Biological Diversity*, League for Pastoral Peoples and Endogenous Livestock Development, Ober-Ramstadt, Germany, http://www.pastoralpeoples.org/wp-content/uploads/2019/11/Community-protocols-web.pdf.
8. 'The Nagoya Protocol on Access and Benefit-Sharing', Convention on Biological Diversity, https://www.cbd.int/abs/infokit/revised/web/factsheet-nagoya-en.pdf.
9. Raika Biocultural Protocol, http://www.pastoralpeoples.org/wp content/uploads/2020/01/Raika_Biocultural_Protocol.pdf.
10. 'Segovia Declaration of Nomadic and Transhumant Pastoralists', WAMIP, World Alliance of Mobile Indigenous Peoples, 14 September, 2007, https://wamipglobal.com/wp-content/uploads/2021/11/Segovia-Pastoralists-Declaration-final.docx-1-1.pdf.
11. 'United Nations Declare 2026 the International Year of Rangelands & Pastoralists', IYRP 2026, https://iyrp.info/sites/iyrp.org/files/Press%20Release%20IYRP%20UNGA%2015.03.22%20with%20logo%202026_0.pdf.

## Appendix 2

1. 'Declaration on Livestock Keepers' Rights', League for Pastoral Peoples, http://www.pastoralpeoples.org/wp-content/uploads/2020/01/Raika_Biocultural_Protocol.pdf.

# Bibliography


Adams, Christina. *Camel Crazy: A Quest for Miracles in the Mysterious World of Camels*. Novato, CA: New World Library, 2019.

Adesogan, Adegbola T. et al. 'Animal Source Foods: Sustainability Problem or Malnutrition and Sustainability Solution? Perspective Matters', *Global Food Security* 25 (2020): 100325. https://doi.org/10.1016/j.gfs.2019.100325.

African Union Department of Rural Economy and Agriculture. *Policy Framework for Pastoralism in Africa: Securing, Protecting and Improving the Lives, Livelihoods and Rights of Pastoralist Communities*. Addis Ababa: African Union, 2010. https://au.int/sites/default/files/documents/30240-doc-policy_framework_for_pastoralism.pdf.

Agrawal, R.P. et al. 'Zero Prevalence of Diabetes in Camel Milk Consuming Raica Community of North-West Rajasthan, India'. *Diabetes Research and Clinical Practice* 76, no. 2 (2007): 290–96. https://doi.org/10.1016/j.diabres.2006.09.036.

Allen, Myles R. et al. 'A Solution to the Misrepresentations of $CO_2$-Equivalent Emissions of Short-Lived Climate Pollutants under Ambitious Mitigation'. *NPJ Climate and Atmospheric Science* 1, no. 16 (2018). https://doi.org/10.1038/s41612-018-0026-8.

Apfelbaum, Steven I. et al. 'Vegetation, Water Infiltration, and Soil Carbon Response to Adaptive Multi-Paddock and Conventional Grazing in Southeastern USA Ranches'. *Journal of Environmental Management* 308 (April 2022): 114576. https://doi.org/10.1016/j.jenvman.2022.114576.

Baldrey, F.S.H. *The Indigenous Breeds of Cattle in Rajputana*. Calcutta: Government Printing, 1909. https://rarebooksocietyofindia.org/book_archive/196174216674_10153579449861675.pdf.

Banning, Edward, and Ilse Köhler-Rollefson. 'Ethnoarchaeological Survey in the Beidha Area, Southern Jordan'. *Annual of the Department of Antiquity of Jordan* 27 (1983): 375–384.

Bassi, Marco. 'Pastoralists Are Peoples: Key Issues in Advocacy and the Emergence of Pastoralists' Rights.' *Nomadic Peoples* 21, no. 1 (2017): 4–33.

Beck, Lois. *The Qashqa''i in an Era of Change: Nomads in Postrevolutionary Iran*. Abingdon: Routledge, 2014.

Behnke, Roy, and Carol Kerven. 'Counting the Costs: Replacing Pastoralism with Irrigated Agriculture in the Awash Valley, North-Eastern Ethiopia'. IIED Climate Change Working Paper, March 2013. https://pubs.iied.org/10035iied.

Behnke, Roy, and Michael Mortimore, eds. *The End of Desertification? Disputing Environmental Change in the Drylands*. Berlin: Springer, 2016.

Benanav, Michael. 'The Sheep Are Like Our Parents'. *New York Times*, 27 July 2012. http://www.nytimes.com/2012/07/29/travel/following-a-navajo-sheep-herder.html.

Benanav, Michael. *Himalaya Bound: One Family's Quest to Save Their Animals – And an Ancient Way of Life*. New York: Pegasus Books, 2018.

Bhattacharya, Neeldri. *The Great Agrarian Conquest: The Colonial Reshaping of a Rural World*. Delhi: Permanent Black, 2018.

Bianchet, Marco Festa. 'Learning to Migrate'. *Science* 361 (2018): 972–97. https://doi.org/10.1126/science.aau6835.

Blanchette, Alex. *Porkopolis: American Animality, Standardized Life, and the Factory Farm*. Durham: Duke University Press, 2020.

Bökönyi, Sandor. 'Archaeological Problems & Methods of Recognizing Animal Domestication'. In *The Domestication and Exploitation of Plants and Animals*, edited by P.J. Ucko and G.W. Dimbleby, 219–229. London: Duckworth, 1969.

Boren, Zach. 'Meat Industry Pushes UN Food Summit to Back Factory Farming'. *Unearthed*, 29 September 2021. https://unearthed.greenpeace.org/2021/09/21/un-food-systems-summit-meat-climate/.

Bradshaw, Corey et al. 'Underestimating the Challenges of Avoiding a Ghastly Future'. *Frontiers in Conservation Science* 1 (2021). https://doi.org/10.3389/fcosc.2020.615419.

Burger, Pamela, E. Ciani and B. Faye. 'Old World Camels in a Modern World – A Balancing Act Between Conservation and Genetic Improvement'. *Animal Genetics* 50 (2019): 598–612. https://doi.org/10.1111/age.12858.

Butaye, Jan, Dries Adriaens and Olivier Honnay. 'Conservation and Restoration of Calcareous Grasslands: A Concise Review of the Effects of Fragmentation and Management on Plant Species'. *Biotechnology, Agronomy, Society, Environment* 9, no. 2 (2005): 111–118.

Cain, Michelle et al. 'Improved Calculation of Warming-Equivalent Emissions for Short-Lived Climate Pollutants'. *NPJ Climate and Atmospheric Science* 2, no. 29 (2019). https://doi.org/10.1038/s41612-019-0086-4.

Cameron, Sarah. 'The Kazakh Famine of 1930–33: Current Research and New Directions'. *East/West Journal of Ukrainian Studies*, no. 2 (2016): 117–132.

Casimir, Michael, 'Of Lions, Herders and Conservationists: Brief Notes on the Gir Forest National Park in Gujarat (Western India)'. *Nomadic Peoples* 5, no. 2, Special Issue: Environment, Property Resources and the State (2001): 154–162.

Chhangani, Anil, Paul Robbins, and S.M. Mohnot, 'Crop Raiding and Livestock Predation at Kumbhalgarh Wildlife Sanctuary, Rajasthan, India', *Human Dimensions of Wildlife* 13 (2008): 305–316. https://doi.org/10.1080/10871200802282922.

Clutton-Brock, Juliet. *Animals as Domesticates: A World View through History*. East Lansing: Michigan State University Press, 2012.

Cracknell, Steve. *The Implausible Rewilding of the Pyrenees*. Lulu.com, 2021.

Cross, H.E. *The Camel and its Diseases: Being Notes for Veterinary Surgeons and Commandants of Camel Corps*. London: Bailliere, Tindall & Cox, 1917.

de Jode, Helen. *The Green Quarter: A Decade of Progress Across the World in Sustainable Pastoralism*. Nairobi: IUCN, 2014. https://www.iucn.org/resources/publication/green-quarter-decade-progress-across-world-sustainable-pastoralism.

DeJohn Anderson, Virginia, *Creatures of Empire: How Domestic Animals Transformed Early America*. Oxford: Oxford University Press, 2004.

Delgado, Christopher, Mark Rosegrant and Henning Steinfeld. 'Livestock to 2020: The Next Food Revolution'. *Outlook on Agriculture* 30, no.1 (2001): 27–29. https://doi.org/10.5367/000000001101293427.

Dessie, Tadelle, and Okore Mwai. *The Story of Cattle in Africa: Why Diversity Matters*. Nairobi, Kenya: International Livestock Research Institute, Rural Development Administration of the Republic of Korea and the African Union-InterAfrican Bureau for Animal Resources, 268, 2019.

Dhongdue, Kyinzom, Gabriel Lafitte and Simon Bradshaw. *An Iron Fist in a Green Glove: Emptying Pastoral Tibet with China's National Parks*. Australia Tibet Council, 2019. https://www.atc.org.au/wp-content/uploads/2019/06/An-Iron-Fist-in-a-Green-Glove_online.pdf.

Diamond, Jared. *Collapse: How Societies Choose to Fail or Succeed*. London: Penguin Books, 2011.

Dongming, Chen et al. 'The Effect of Different Restoration Measures on the Desertified Alpine Grassland in Zoigê'. *Chinese Journal of Applied & Environmental Biology* 22 (2016): 573–578.

Dransart, Penelope. *Earth, Water, Fleece and Fabric: An Ethnography and Archaeology of Andean Camelid Herding*. London: Routledge, 2002.

Ekvall, Robert. *Fields on the Hoof: Nexus of Tibetan Nomadic Pastoralism*. New York: Holt, Rinehart and Winston, 1968.

Elmi, Ahmed. *Camel Husbandry and Management by Celdheer Pastoralists in Central Somalia*. Pastoral Development Network Paper 27d. London: Overseas Development Institute, 1989.

Eriksson, Camilla, 'What is Traditional Pastoral Farming? The Politics of Heritage and "Real Values" in Swedish Summer Farms (Fäbodbruk)'. *Pastoralism* 1, no. 25 (2011). https://doi.org/10.1186/2041-7136-1-25.

European Commission, 'Improving the Management of Salisbury Plain Natura 2000 Sites'. https://webgate.ec.europa.eu/life/publicWebsite/index.cfm?fuseaction=search.dspPage&n_proj_id=1712 (accessed 24 April 2022).

Fan, Mingming et al. 'Impacts of Nomad Sedentarization on Social and Ecological Systems at Multiple Scales in Xinjiang Uyghur Autonomous Region, China'. *Ambio* 43, no. 5 (2014): 673–86. https://doi.org/10.1007/s13280-013-0445-z.

Fernandez-Gimenez, Maria E. 'The Role of Mongolian Nomadic Pastoralists' Ecological Knowledge in Rangeland Management'. *Ecological Applications* 10, no. 5 (2000): 1318–1326. http://www.jstor.org/stable/2641287.

Fijn, Natasha. *Living with Herds: Human-Animal Co-Existence in Mongolia*. Cambridge: Cambridge University Press, 2011.

Fischer, Sabine, Peter Poschlod and Burkhard Beinlich. 'Experimental Studies on the Dispersal of Plants and Animals on Sheep in Calcareous Grasslands'. *Journal of Applied Ecology* 33, no. 65 (1996): 1206–1222. https://doi.org/10.2307/2404699.

Food and Agriculture Organization of the United Nations. *Global Plan of Action for Animal Genetic Resources and the Interlaken Declaration*. Rome: FAO, 2007. https://www.fao.org/3/a1404e/a1404e.pdf.

Food and Agriculture Organization of the United Nations. *Livestock's Long Shadow: Environmental Issues and Options*. Rome: FAO, 2006. https://www.fao.org/3/a0701e/a0701e.pdf.

Food and Agriculture Organization of the United Nations. *Pastoralism – Making Variability Work*. FAO Animal Production and Health Paper No. 185. Rome: FAO, 2021. https://doi.org/10.4060/cb5855en.

Food and Agriculture Organization of the United Nations. *The State of the World's Biodiversity for Food and Agriculture*, edited by Julie Bélanger and Dafydd Pilling (FAO Commission on Genetic Resources for Food and Agriculture Assessments: Rome, 2019). http://www.fao.org/3/CA3129EN/CA3129EN.pdf.

Fryxell, John M., and Anthony Sinclair. 'Causes and Consequences of Migration by Large Herbivores'. *Trends in Ecology and Evolution* 3 (1988): 237–241.

Galaty, John. 'Cattle and Cognition: Aspects of Maasai Practical Reasoning'. In *The Walking Larder: Patterns of Domestication, Pastoralism, and Predation*, edited by Juliet Clutton-Brock, 215–230. London: Unwin Hyman, 1989.

Garnett, Tara et al. *Grazed and Confused? Ruminating on Cattle, Grazing Systems, Methane, Nitrous Oxide, The Soil Carbon Sequestration Question – and What It All Means for Greenhouse Gas Emissions*. Oxford: FCRN, University of Oxford, 2017. https://www.oxfordmartin.ox.ac.uk/downloads/reports/fcrn_gnc_report.pdf.

Gauthier-Pilters, Hilde, and Anne I. Dagg. *The Camel: Its Evolution, Ecology, Behaviour and Relationship to Man*. Chicago: Chicago University Press, 1981.

Geerlings, Ellen. 'The Black Sheep of Rajasthan'. *Seedling* (2004). https://grain.org/article/entries/436-the-black-sheep-of-rajasthan.

Geng, Y. et al. 'Prioritizing Fodder Species Based on Traditional Knowledge: A Case Study of Mithun (Bos frontalis) in Dulongjiang area, Yunnan Province, Southwest China'. *Journal of Ethnobiology and Ethnomedicine* 13, no. 24 (2017). https://doi.org/10.1186/s13002-017-0153-z.

Gerber, Pierre et al. *Tackling Climate Change through Livestock – A Global Assessment of Emissions and Mitigation Opportunities*. Rome: Food and Agriculture Organization of the United Nations, 2013.

GRAIN, 'Emissions Impossible: How Big Meat and Dairy are Heating up the Planet', 18 July 2018. https://grain.org/article/entries/5976-emissions-impossible-how-big-meat-and-dairy-are-heating-up-the-planet.

Häckh, Ruth. *Eine für Alle: Mein Leben als Schäferin*. München: Ludwig Buchverlag, 2018.

Hartley, Brian. 'The Dromedary of the Horn of Africa'. In *The Camelid: An All-Purpose Animal, Vol. I.*, edited by W.R. Cockrill, 77–97. Uppsala: Scandinavian Institute of African Studies, 1984.

Havlík, Petr et al. 'Climate Change Mitigation through Livestock System Transitions'. *Proceedings of the National Academy of Sciences* 111, no 10 (2014): 3709–3714. https://doi.org/10.1073/pnas.1308044111.

Heitschmidt, Rod, R.E. Short and E.E. Grings. 'Ecosystems, Sustainability, and Animal Agriculture'. *Journal of Animal Science* 74, no. 6 (1996): 1395–405. https://doi.org/10.2527/1996.7461395x.

Hertkorn, Marie-Luise, Hassan Roba and Brigitte Kaufmann. 'Caring for Livestock: Borana Women's Perceptions of Their Changing Role in Livestock Management in Southern Ethiopia'. *Nomadic Peoples* 19, no. 1 (2015): 30–52.

Hjort, Anders, and Gudrun Dahl. 'A Note on the Camel of the Amar'ar Beja'. In *The Camelid: An All-Purpose Animal, Vol I.*, edited by W.R. Cockrill, 50–76. Uppsala: Scandinavian Institute of African Studies (1984).

Hjort, Anders, and Gudrun Dahl. *Having Herds: Pastoral Herd Growth and Household Economy*. Stockholm: Department of Social Anthropology, Stockholm University, 1984: 52–53.

Houzer, Ella, and Ian Scoones. *Are Livestock Always Bad for the Planet? Rethinking the Protein Transition and Climate Change Debate*. Brighton: Institute of Development Studies, 2021. https://pastres.org/livestock-report/.

Howard, Albert. 'The Animal as Our Farming Partner'. *Organic Gardening* 2, no. 3 (1947). http://journeytoforever.org/farm_library/howard_animal.html.

Hristov, Anthony N. 'Historic, Pre-European and Present-Day Contribution of Wild Ruminants to Enteric Methane Emissions in the United States'. *Journal of Animal Science* 90 (2012): 1371–1375. https://pubmed.ncbi.nlm.nih.gov/22178852/.

Hülsebusch, Christian, Brigitte Kaufmann and Marion Adams. *Camel Breeds and Breeding in Northern Kenya*. Nairobi: Kenya Agricultural Research Institute, 2002.

Hutchins, Kip. 'Like a Lullaby: Song as Herding Tool in Rural Mongolia'. *Journal of Ethnobiology* 39 no. 3 (2019): 445–459. https://doi.org/10.2993/0278-0771-39.3.445.

Hyun Jin Kim. 'The Xiongnu'. *Asian History, Oxford Research Encyclopedias*, 2017. https://doi.org/10.1093/acrefore/9780190277727.013.50.

International Museum of the Horse. *A Gift from the Desert: The Art, History and Culture of the Arabian Horse*. 2010. https://archive.org/details/1GFDCatalogFront/page/n17/mode/2up.

Ivarsdotter, Anna. 'And the Cattle Follow Her, for They Know Her Voice… Of Communication Between Women and Cattle in Scandinavian Pastures'. In *PECUS Man and animals in antiquity: Proceedings of the conference at the Swedish Institute in Rome,* 9–12 September 2002, edited by Barbro Santillo Frizell, 146–149. Rome: The Swedish Institute in Rome, 2004.

Iverson, Peter. *Diné: A History of the Navajos*. Albuquerque, NM: University of New Mexico Press, 2002.

Jesmer, Brett et al. 'Is Ungulate Migration Culturally Transmitted? Evidence of Social Learning from Translocated Animals'. *Science* 361, no. 6406 (2018): 1023–1025. https://doi.org/10.1126/science.aat0985.

Joshi, Srijana, et al. 'Ethnic and Cultural Diversity amongst Yak Herding Communities in the Asian Highlands'. *Sustainability* 12 (2020): 957. https://doi.org/10.3390/su12030957.

Klute, Georg. *Die Schwerste Arbeit der Welt. Alltag von Tuareg Nomaden*. Wuppertal, Germany: Trickster Verlag, 1992.

Köhler-Rollefson, Ilse. 'Between Burning Irons and Antibiotics'. *Reports of the DFG* 2–3 (1997): 4–6.

Köhler-Rollefson, Ilse. 'Changes in Goat Exploitation at 'Ain Ghazal between the Early and Late Neolithic: A Metrical Analysis'. *Paléorient* 15 no. 1 (1989): 141–146. https://doi.org/10.3406/paleo.1989.4492.

Köhler-Rollefson, Ilse. 'Indigenous Practices of Animal Genetic Resource Management and Their Relevance for the Conservation of Domestic Animal Diversity in Developing Countries'. *Journal of Animal Breeding and Genetics* 114 (1997): 231–238. https://pubmed.ncbi. nlm.nih.gov/21395819/.

Köhler-Rollefson, Ilse, and Gary Rollefson. 'Brooding about Breeding: Social Implications for the Process of Animal Domestication'. In *The Dawn of Farming in the Near East*, edited by R.T.J. Cappers and S. Bottema, 177–182. Berlin: Ex Oriente, 2002.

Köhler-Rollefson, Ilse, and Hanwant Singh Rathore. 'Indigenous versus Official Knowledge, Concepts and Institutions: Raika Pastoralists and the Outside World'. *Nomadic Peoples* 8, no. 2 (2004): 150–167. https://doi.org/10.3167/082279404780446104.

Köhler-Rollefson, Ilse, and Hanwant Singh Rathore. 'The Case of the Kumbhalgarh Wildlife Sanctuary and Camel Pastoralism in Rajasthan (India)'. *Sustainability* 13, no. 24) (2021): 13914. https://doi.org/10.3390/su132413914.

Koocheki, Alireza. 'Herders Care for Their Land'. *ILEIA Newsletter* 8, no. 3 (1992): 3.

Kotschi, Johannes. *A Soiled Reputation: Adverse Impacts of Mineral Fertilizers in Tropical Agriculture*. Berlin: Heinrich Böll Foundation and WWF, Germany, 2013.

Krätli, Saverio. 'Cattle Breeding, Complexity and Mobility in a Structurally Unpredictable Environment: The WoDaaBe Herders of Niger'. *Nomadic Peoples* 12, no. 1 (2008): 11–41.

Krätli, Saverio. 'Cows Who Choose Domestication: Generation and Management of Domestic Animal Diversity by WoDaaBe Pastoralists (Niger)'. PhD dissertation, Institute of Development Studies at the University of Sussex U.K., 2007.

Krätli, Saverio et al. 'Pastoralism: A Critical Asset for Food Security under Global Climate Change'. *Animal Frontiers* 3, no. 1 (January 2013): 42–50. https://doi.org/10.2527/af.2013-0007.

Lainé, Nicolas, and Serge Morand. 'Linking Humans, Their Animals, and the Environment Again: A Decolonized and More-than-Human Approach to "One Health"'. *Parasite* 27 (2020): 55. https://doi.org/10.1051/parasite/2020055.

League for Pastoral Peoples. *The Karen Commitment: Proceedings of a Conference of Indigenous Livestock Breeding Communities on Animal Genetic* Resources, 2020. http://www.pastoralpeoples.org/wp-content/uploads/2020/01/karen1.pdf.

Lewis, I.M. *A Pastoral Democracy: A Study of Pastoralism and Politics among the Northern Somali of the Horn of Africa*. Oxford: Oxford University Press, 1961: 85.

Lewis, Michael. 'Cattle and Conservation at Bharatpur: A Case Study in Science and Advocacy'. *Conservation and Society* 1(1) (2003): 1–21. http://www.jstor.org/stable/26396448.

Liboiron, Max. 'Decolonizing Geoscience Requires More Than Equity and Inclusion'. *Nature Geosciences* 14 (2021): 876–877. https://doi.org/10.1038/s41561-021-00861-7.

Logsdon, Gene. *Holy Shit: Managing Manure to Save Mankind*. White River Junction: Chelsea Green, 2010.

LPPS. *The Camels of Kumbhalgarh: A Biodiversity Treasure*. Sadri, India: LPPS, 2013. http://www.lpps.org/wp-content/uploads/2013/10/Camels_Of_Kumbhalgarh_web.pdf.

Majgaonkar, Iravatee et al. 'Land-sharing Potential of Large Carnivores in Human-Modified Landscapes of Western India'. *Conservation Science and Practice* 1 no. 5 (2019): e34. https://doi.org/10.1111/csp2.34.

Manzano, Pablo, and Juan Malo. 'Extreme Long-Distance Seed Dispersal via Sheep'. *Frontiers in Ecology and Environment* 4, no. 5 (2006): 244–248. https://www.jstor.org/stable/3868790 .

Marshall, Fiona et al. 'Ancient Herders Enriched and Restructured African Grasslands'. *Nature* 561 (7723) (2018): 387–390. https://doi.org/10.1038/s41586-018-0456-9.

Mekonnen, Mesfin, and Arjen Hoekstra. 'A Global Assessment of the Water Footprint of Farm Animals and Animal Products'. *Ecosystems* 15, no. 3 (2012): 401–415. https://doi.org/10.1007/s10021-011-9517-8.

Meuret, Michel et al. *The Art and Science of Shepherding: Tapping the Wisdom of French Herders*. Greeley, Colorado: Acres USA, 2014.

Milman, Oliver. 'Five Hundred Goats Save the Ronald Reagan Library from Wildfires'. *The Guardian*, 31 October 2019. https://www.theguardian.com/us-news/2019/oct/31/goats-save-ronald-reagan-library-wildfire.

Musil, Alois. *The Manners and Customs of the Rwala Bedouins*. New York: The American Geographical Society, 1928.

Niamir-Fuller, M., C. Kerven, R. Reid and E. Milnar-Gulland. 'Co-Existence of Wildlife and Pastoralism on Extensive Rangelands: Competition or Compatibility?'. *Pastoralism* 2, no. 8 (2012). https://doi.org/10.1186/2041-7136-2-8.

Nickel, Herbert. 'Zikaden'. In M. Bunzel-Drüke et al., *Naturnahe Beweidung und Natura2000*, 267–277. Bad Sassendorf (Germany): Arbeitsgemeinschaft Biologischer Umweltschutz, 2019.

Nuttal, Philippa. 'Pat Brown: "Farm Animals Are the Most Destructive Technology on Earth"'. *New Statesman*, last updated 2 December 2021. https://www.newstatesman.com/the-environment-interview/2021/11/pat-brown-farm-animals-are-the-most-destructive-technology-on-earth.

Nyariki, D.M., and D.A Amwata. 'The Value of Pastoralism in Kenya: Application of Total Economic Value Approach'. *Pastoralism* 9, no. 9 (2019). https://doi.org/10.1186/s13570-019-0144-x.

Obua, Be. 'Checklist of Forage Plants Utilized for Sheep Feeding in Nsukka Area of Enugu State, Southeastern Nigeria'. *National Journal of Agriculture and Rural Development* 2, no. 2 (2018): 3548–3565.

Obua, Be. 'Survey of the Diversity of Browse Plants Utilized for Goat Feeding in Ohaji/Egbema/Oguta Area of Imo State, Nigeria'. *International Journal of Tropical Agriculture and Food Systems* 7, no. 1 (2013): 54–66.

Oldenbroek. Kor, and Liesbeth van der Waaij. *Textbook Animal Breeding: Animal Breeding and Genetics for BSc students*. Wageningen: Wageningen University and Research Centre, the Netherlands, 2014. https://www.wur.nl/upload_mm/4/d/5/3b256d1f-2ae5-4fa6-a2d7-afe355ce9870_Textbook%20Animal%20Breeding%20and%20Genetics-v17-20160105_2020.pdf.

Oleaa, Pedro P., and Patricia Mateo-Tomás. 'The Role of Traditional Farming Practices in Ecosystem Conservation: The Case of Transhumance and Vultures', *Biological Conservation* 142 (2009): 1844–1853. https://doi.org/10.1016/j.biocon.2009.03.024.

Ostrom, Elinor et al, eds, *The Drama of the Commons*. Washington, D.C.: National Academies Press, 2002.

Pardikar, Rishika. 'Large Herbivores May Improve an Ecosystem's Carbon Persistence'. *Eos* 103 (2022): https://doi.org/10.1029/2022EO220029.

Park, S. et al. 'Trends and Seasonal Cycles in the Isotopic Composition of Nitrous Oxide Since 1940'. *Nature Geoscience* 5 (2012): 261–265. https://doi.org/10.1038/ngeo1421.

Podger, Pamela. 'Got Weeds? These Sheep Will Make House Calls'. *New York Times*, 26 October 2008. https://www.nytimes.com/2008/10/27/us/27weeds.html.

Poore, Joseph, and Thomas Nemecek. 'Reducing Food's Environmental Impacts Through Producers and Consumers'. *Science* 360 (2018): 987–992. https://doi.org/10.1126/science.aaq0216.

Porcher, Jocelyne. 'The Work of Animals: A Challenge for Social Sciences'. *Humanimalias* 6, no. 1 (2014): https://humanimalia.org/article/view/9925.

Porter, Roy. *The Penguin Social History of Britain: English Society in the Eighteenth Century*. London: Penguin, 1982.

Pörtner, Hans-Otto, et al. *IPBES-IPCC Co-Sponsored Workshop Report on Biodiversity and Climate Change*, IPBES and IPCC, 24 June 2021. https://zenodo.org/record/5101133#.YpMQOUDhWM8.

Poschlod, Peter. 'The Historical and Socioeconomic Perspective of Calcareous Grasslands–Lessons from the Distant and Recent Past'. *Biological Conservation* 104, no. 3 (2002): 361–376. https://doi.org/10.1016/S0006-3207(01)00201-4.

Provenza, Fred, Michel Meuret and Pablo Gregorini. 'Our Landscapes, Our Livestock, Ourselves: Restoring Broken Linkages among Plants, Herbivores, and Humans with Diets that Nourish and Satiate'. *Appetite* (December 2015): 500–519. https://doi.org/10.1016/j.appet.2015.08.004.

Purohit, Kush, Hanwant Singh Rathore and Ilse Köhler-Rollefson. 'Increased Risk of Type 2 Diabetes Mellitus in the Maru Raika Community of Rajasthan: A Cross-Sectional Study'. *International Journal of Diabetes in Developing Countries* 37 (2016): 494–501. http://dx.doi.org/10.1007/s13410-016-0529-y.

Ragkos, Athanasios et al. 'Current Trends in the Transhumant Cattle Sector in Greece'. *Scientific Papers: Animal Science and Biotechnologies* 46, no. 1 (2013).

Rao, Mohit M. 'Why There's no Conflict Between Wolves and Shepherds in These Koppal Villages'. *The Hindu*, 22 December 2018. https://www.thehindu.com/sci-tech/energy-and-environment/why-theres-no-conflict-between-wolves-and-shepherds-in-these-koppal-villages/article25789757.ece.

Riesman, Paul. *Freedom in Fulani Social Life: An Introspective Ethnography*. Chicago: University of Chicago Press, 1977, 102.

Ritchie, Hannah, and Max Roser. 'Biodiversity'. Published online at OurWorldInData.org. Retrieved from https://ourworldindata.org/biodiversity [Online Resource] and accessed on 13 February 2022.

Roche, Gerald. 'Abandoning the High Ground: The Ecological Implications of Pastoral Abandonment in Tibet'. Presented at the 3rd Himalayan Studies Conference, Yale, 14–16 March 2014.

Roe, Emery, Lyn Huntsinger and K. Labnow. 'High-Reliability Pastoralism Versus Risk-Averse Pastoralism'. *Journal of Environment and Development* 7, no. 4 (1998): 387–421.

Roe, Emery. *A New Policy Narrative for Pastoralism? Pastoralists as Reliability Professionals and Pastoralist Systems as Infrastructure*. STEPS

Working Paper No. 113. Brighton, U.K.: Social, Technological and Environmental Pathways to Sustainability (STEPS) Centre, Institute of Development Studies, 2020.

Rolandsen, Christer M. et al. 'On Fitness and Partial Migration in a Large Herbivore – Migratory Moose Have Higher Reproductive Performance Than Residents'. *Oikos* 126 (4) (2017): 547–555.

Salehi, Bahare et al. 'Antidiabetic Potential of Medicinal Plants and Their Active Components'. *Biomolecules* 9, no. 10 (2019): 551. https://doi.org/10.3390/biom9100551.

Sato, Shun. 'Pastoral Movements and the Subsistence Unit of the Rendille of Northern Kenya: with Special Reference to Camel Ecology'. *Senri Ethnological Studies* 6 (1980).

Sayre, Nathan. *The Politics of Scale: A History of Rangeland Science*. Chicago: University of Chicago Press, 2017.

Schareika, Nikolaus, Christopher Brown and Mark Moritz. 'Critical Transitions from Pastoralism to Ranching in Central Africa'. *Current Anthropology* 62, no.1 (2021): 53–76.

Schlee, Günther. 'Camel Management Strategies and Attitudes Towards Camels in the Horn'. In *The Exploitation of Animals in Africa*, edited by J. Stone, 143–154. Aberdeen: Aberdeen University, African Studies Group, 1989.

Seid, M.A., N.J. Kuhn and T.Z. Fikre. 'The Role of Pastoralism in Regulating Ecosystem Services'. *Revue Scientifique et Technique (International Office of Epizootics)* 35 no.2 (2016): 435–444. https://doi.org/10.20506/rst.35.2.2534.

Sharma, Anita. *The Bakkarwals of Jammu and Kashmir: Navigating through Nomadism*. New-Delhi: Niyogi Books, 2009.

Shiva, Vandana. 'In Praise of Cowdung'. Znet, 2002. https://zcomm.org/znetarticle/in-praise-of-cowdung-by-vandana2-shiva-1/.

Simmons, Alan et al. 'Ain Ghazal: A Major Neolithic Settlement in Central Jordan', *Science* 240, 4848 (1988): 35–39. https://doi.org/10.1126/science.240.4848.35.

Smith, John Masson Jr, 'Review: Nomads on Ponies vs. Slaves on Horses', *Journal of the American Oriental Society* 118, no. 1 (1998): 54–62. https://doi.org/10.2307/606298

Spate, Michael et al. 'Palaeoenvironmental Proxies Indicate Long-Term Development of Agro-Pastoralist Landscapes in Inner Asian

Mountains'. *Scientific Reports* 12 (2022): 554. https://www.nature.com/articles/s41598-021-04546-4.

Stammler, Florian. 'Animal Diversity and its Social Significance Among Arctic Pastoralists'. In *Good to Eat, Good to Live with: Nomads and Animals in Northern Eurasia and Africa*, edited by Florian Stammler and Hiroki Takakura. Sendai, Japan: Center for Northeast Asian Studies, Tohoku University, 2010.

Starrs, Paul F. 'Transhumance as Antidote for Modern Sedentary Stock Raising'. *Rangeland Ecology & Management* 71 no. 5 (2018): 592–602. https://doi.org/10.1016/j.rama.2018.04.011.

Steinfeld, Henning, Pierre Gerber and Carolyn Opio. 'Responses on Environmental Issues'. In *Livestock in a Changing Landscape: Drivers, Consequences, and Responses*, edited by Henning Steinfeld et al. Washington, D.C.: Island Press, 2010, 313.

Stépanoff, Charles et al. 'Animal Autonomy and Intermittent Coexistences: North Asian Modes of Herding'. *Current Anthropology* 58 no. 1 (February 2017): 57–81. https://www.journals.uchicago.edu/doi/full/10.1086/690120.

Teague, W.R. et al. 'The Role of Ruminants in Reducing Agriculture's Carbon Footprint in North America'. *Journal of Soil and Water Conservation* 71, no. 2 (2016): 156–164. https://doi.org/10.2489/jswc.71.2.156.

Teka, Tegegne. 'Camel and the Household Economy of the Afar'. *Nomadic Peoples* 29 (1991): 31–41. https://www.jstor.org/stable/43123336.

Thomas, D.T. et al. 'Net Protein Contribution and Enteric Methane Production of Pasture and Grain-Finished Beef Cattle Supply Chains'. *Animal* 5 no.12 (2021): 100392. https://doi.org/10.1016/j.animal.2021.100392.

Tibetan Centre for Human Rights & Democracy. *Wasted Lives: A Critical Analysis of China's Campaign to End Tibetan Pastoral Lifeways*, 2015. https://tchrd.org/wasted-lives-new-report-offer-fresh-insights-on-travails-of-tibetan-nomads/.

Uerpmann, Hans-Peter. 'Animal Domestication ± Accident or Intention?'. In *The Origin and Spread of Agriculture and Pastoralism in Eurasia*, edited by David Harris, 227–237. London: University College London Press, 1996.

UNEP. 'Back to the Future: Rangeland Management in Jordan'. UNEP News and Stories, 22 June 2016. https://www.unenvironment.org/news-and-stories/story/back-future-rangeland-management-jordan.

Van Vliet, Stephan, Frederick D. Provenza and Scott L. Kronberg. 'Health-Promoting Phytonutrients Are Higher in Grass-Fed Meat and Milk'. *Frontiers of Sustainable Food Systems*, 2 February 2021. https://doi.org/10.3389/fsufs.2020.555426.

Van Zanten, Hannah et al. 'Defining a Land Boundary for Sustainable Livestock Consumption'. *Global Change Biology* 9 (2018): 4185–4194. https://doi.org/10.1111/gcb.14321.

Van Zanten, Hannah, Martin K. Van Ittersum and Imke J.M. De Boera. 'The Role of Farm Animals in a Circular Food System'. *Global Food Security* 21 (2019):18–22. https://doi.org/10.1016/j.gfs.2019.06.003.

Vásquez, Grimaldo. 'Culture and Biodiversity in the Andes'. *Forest, Trees and People Newsletter* 34 (1997): 39–45.

Ventresca Miller, Alicia et al. 'Ecosystem Engineering Among Ancient Pastoralists in Northern Central Asia'. *Frontiers Earth Science* 8 (2020): 168. https://doi.org/10.3389/feart.2020.00168.

Vera, Frans. *Grazing Ecology and Forest History*. Wallingford, U.K.: CABI Publishing, 2000.

Verma, Megha et al. 'Can Reindeer Husbandry Management Slow Down the Shrubification of the Arctic?'. *Journal of Environmental Management* 267:110636 (2020). https://doi.org/10.1016/j.jenvman.2020.110636.

Vidal, John. 'Health Risks of Shipping Pollution Have Been Underestimated'. *The Guardian*, 9 April 2009. https://www.theguardian.com/environment/2009/apr/09/shipping-pollution.

Vitebsky, Piers. *Reindeer People: Living with Animals and Spirits in Siberia*. London: HarperCollins, 2005.

Vitebsky, Piers. 'Wild Tungus and the Spirits of Places'. *Ab Imperio* 2012, no. 2 (2012): 429–448. https://doi.org/10.1353/imp.2012.0046.

Weisiger, Marsha, *Dreaming of Sheep in Navajo Country*. Seattle: University of Washington Press, 2011.

Whitehead, John. 'John Locke and the Governance of India's Landscape: The Category of Wasteland in Colonial Revenue and Forest Legislation'. *Economic and Political Weekly* 45, no. 50 (2010): 83–93.

Zogib, Lisa. *On the Move – for 10,000 years: Biodiversity Conservation Through Transhumance and Nomadic Pastoralism in the Mediterranean*. The Mediterranean Consortium for Nature and Culture, 2014. https://www.mednatureculture.org/wp-content/uploads/2021/08/10000Years_MediterraneanConsortiumForNatureAndCulture.pdf.

# Index

**D**

T



U



V



W

# About the Author

HANWANT SINGH RATHORE

Ilse Köhler-Rollefson studied veterinary medicine in Germany before working as an archaeozoologist in Jordan where she discovered her fascination for camels and herding cultures. After completing her PhD on camel domestication, she studied the Raika camel culture of India which led her to found the League for Pastoral Peoples (www.pastoralpeoples.org), an international advocacy organisation that is giving a voice to herders at the global level.

Having authored well over a hundred scientific publications, she consults with and writes for the UN and other development agencies. She is the author of a memoir, *Camel Karma: Twenty Years Among India's Camel Nomads,* as well as co-editor of the *Field Manual on Camel Diseases*.

Ilse lives in Rajasthan in India where she owns a small herd of camels and has co-founded the country's first camel dairy. Her work has been recognised by the Maharaja of Jodhpur and she has received India's highest award for women from its president as well as the Order of Merit from the President of Germany. Ilse is regularly quoted and interviewed by mainstream media, including the BBC, *Forbes India* and the *Hindustan Times*, for her expertise in camels, pastoralism and livestock ecology. She has given a TEDX *Talk about The Nomads that Feed Us*, has recently moderated a panel about pastoralism at the High-Level Political Forum of the UN and blogs about 'Livestock Futures' at www.ilse-koehler-rollefson.com.

Twitter @IlseKohler